AF333769

Peter Schuster
Werner Mikenda (eds.)

Hydrogen Bond Research

SpringerWienNewYork

Prof. Dr. Peter Schuster
Institute of Theoretical Chemistry and Molecular Structural Biology,
University of Vienna, Austria

Prof. Dr. Werner Mikenda
Institute of Organic Chemistry, University of Vienna, Austria

This work is subject to copyright.
All rights are reserved, whether the whole or part of the material is concerned, specifically those of translation, reprinting, re-use of illustrations, broadcasting, reproduction by photocopying machines or similar means, and storage in data banks.

© 1999 Springer-Verlag Wien
Printed in Austria

Typesetting: Thomson Press (India) Ltd., New Delhi
Printing: Eugen Ketterl Gesellschaft m.b.H., A-3001 Mauerbach
Printed on acid-free and chlorine-free bleached paper

With 44 Figures

CIP data applied for

Special Edition of *Monatshefte für Chemie/Chemical Monthly*, Vol. 130, No. 8, 1999

ISBN 3-211-83396-X Springer-Verlag Wien New York

Contents

Editorial

The notion of the hydrogen bond has been invented at the beginning of our century and was used by *Latimer* and *Rodebush* in 1920, and independently by *Huggins* in 1922, to characterize a structural regularity in which a hydrogen atom is bound to two neighbors. Almost forgotten in the following 15 years, the hydrogen bond was brought again into the focus of interest when it was 'rediscovered' as a regularity in crystal structures. The development of efficient computational methods in quantum chemistry provided new tools for the analysis of intermolecular forces. This novel approach was used extensively in the seventies to develop a heuristic theoretical concept for the hydrogen bond which allowed to derive quantitative expressions for different contributions, like the electrostatic energy, exchange repulsion energy, polarization energy, and dispersion contribution or charge transfer energy. In principle, charge transfer can also be interpreted as a covalent contribution. Later on very successful developments of computational methods yielded results that were in qualitative agreement with the existing concept and the available experimental data. More recently, it became possible to reach an unforeseeable accuracy in *ab intio* calculations that laid down the basis for successful competition with experimental results from vapor phase spectroscopy. Progress in the experimental techniques for the study of intermolecular complexes, in particular small hydrogen-bonded clusters, was made through molecular beam electronic resonance spectroscopy and application of infrared laser spectroscopy.

In recent years, the major impact in the understanding of hydrogen bonds came from studies on clusters of few to many molecules. The primary issue of investigations on these systems is to obtain precise information of the nature and the extent of deviations from pairwise additivity that is often addressed as hydrogen bond cooperativity. Another problem that had intrigued theorists for a long time is the stability of intramolecular hydrogen bonds. Coupling of these hydrogen bonds to π-electrons, in particular to conjugated π-electron systems, increases substantially their stability, whereas no major effects are observed in absence of conjugation partners. Progress in computational capacities and experimental techniques provided tools that nowadays allow to study and to analyze these subtle phenomena with sufficient reliability.

This issue of *Chemical Monthly* presents a collection of review articles and original papers devoted to hydrogen bonding. Without the claim of full coverage they also contain a representative overview of the research work done in this area at the Austrian universities.

The review of *Schuster* and *Wolschann* aims at a presentation of the state of the art in hydrogen bonding with the ultimate goal to understand the hydrogen bonds in biopolymers.

NMR spectroscopy is one of the most powerful tools to detect and to investigate hydrogen bonds. Recent developments in the methodology of NMR spectroscopy allow to study hydrogen bonding in aqueous solution. The review by *Konrat, Tollinger, Kontaxis*, and *Kräutler* describes the new techniques and presents interesting examples, thus demonstrating the power of the approach.

In a class of *Mannich* bases derived from substituted phenols or naphthols, stable hydrogen bonds are formed in a six-membered ring containing an sp^3 hybridized carbon atom. These rings contain a hydrogen bond without coupling to conjugated π-electrons and hence represent excellent model systems for case studies on intramolecular hydrogen bonding (*Koll, Wolschann*).

Simperler and *Mikenda* study competition between intramolecular hydrogen bonds in substituted phenols with carbonyl groups in positions 2 and 6.

Cyanodiacetylene can adopt two modes of intermolecular interaction in the dimer: hydrogen bonding or antiparallel orientation with a compensation of the dipole moments. The contribution by *Karpfen* shows that accurate calculations clearly favor the latter arrangement.

The paper by *Wolf et al.* presents a molecular dynamics study on proton transfer in coupled pairs of hydrogen bonds. Two representative examples are chosen: the formic acid dimer and 5,8-dihydroxy-1,4-naphthoquinone. The essential difference between the two cases concerns the double proton transfer which is a concerted process in the carboxylic acid dimer, but a consecutive two-step process in the second example.

The paper by *Libowitzky* provides a comparison of OH$\cdots$O bond lengths and OH stretching frequencies in a set of data derived from 65 minerals. The correlation curve shows a strong dependence for short hydrogen bonds, smoothly converging into a constant vibrational frequency for hydrogen bonds with an OO-distance of 3.0 Å or longer.

Without encouragement, patience, and help of Professors *Falk* and *Kalchhauser* this special volume would never have been finished in time. Many thanks to them!

Peter Schuster
Werner Mikenda
Issue Editors

Hydrogen Bonding: From Small Clusters to Biopolymers

Peter Schuster[*] and **Peter Wolschann**

Institut für Theoretische Chemie und Molekulare Strukturbiologie, Universität Wien, A-1090 Wien, Austria

Summary. High quality *ab initio* computations and molecular spectroscopy of small hydrogen-bonded clusters in the vapor phase provide highly accurate data in general agreement with the theory of hydrogen bonds developed in the seventies. Hydrogen bonding is a major force determining energetics and structures of biopolymers. In addition to direct influence through their directionality, hydrogen bonds set the stage for the formation of biopolymer structures indirectly since they determine the water structure. On the basis of current results hydrophobic interactions are considered equally important or even more relevant than direct hydrogen bonding. A new concept for protein and nucleic acid folding which is based on statistical mechanics allows to study the role of hydrogen bond formation in the nucleation process as well as in later states.

Keywords. Cluster; Hydrogen bond; Protein; Nucleic acid; Template.

Wasserstoffbrücken: Von kleinen Clustern zu Biopolymeren

Zusammenfassung. *Ab initio*-Rechnungen von hoher Qualität und die Gasphasenmolekülspektroskopie von kleinen Komplexen mit Wasserstoffbrücken liefern Daten von höchster Genauigkeit, welche in guter Übereinstimmung mit der in den Siebzigerjahren entwickelten Theorie der Wasserstoffbrückenbindung stehen. Wasserstoffbrücken bilden eines der wichtigsten energetischen und stereochemischen Prinzipien, welche die Strukturen der Biopolymeren bestimmen. Zusätzlich zum direkten Einfluß über ihre direktive Wirkung beeinflussen Wasserstoffbrücken aber auch indirekt die Ausbildung von Biopolymerstrukturen, da sie essentiell an der Struktur des flüssigen Wassers mitwirken. Die gegenwärtig zur Verfügung stehenden Daten legen nahe, daß die hydrophobe Wechselwirkung zumindest ebenso wichtig wenn nicht sogar noch wesentlicher ist als die direkten Wasserstoffbrücken. Ein neues Konzept zur Beschreibung der Faltung von Proteinen und Nukleinsäuren auf der Basis der statistischen Mechanik ermöglicht es, die Rolle der Ausbildung von Wasserstoffbrücken auch bei der Nukleation der Faltung und in späteren Phasen des Prozesses zu untersuchen.

---

[*] Corresponding author

1. Introduction

The molecular structures of the two most important classes of biopolymers, proteins and nucleic acids, are largely determined by hydrogen bonds: directly since they are important elements of biopolymer structure, and indirectly through hydrophobic interactions. Historically, the dominant role of hydrogen bonding became apparent in the early fifties through a few publications of model structures for proteins [1] and nucleic acids [2], respectively. The structures of both classes of biopolymers, *i.e.* α-helix and β-sheet of polypeptides and double helices of polynucleotides, are evidently built around the directions defined by optimal hydrogen bond geometry. Today, almost half a century after these milestone discoveries that initiated molecular biology, the understanding of biomolecular structures has become more subtle. The major driving force in structure formation of biopolymers appears to be hydrophobic interaction rather than hydrogen bonding. Helical structures of polypeptides are also formed when there is no possibility to form hydrogen bonds like, for example, in polyproline. Free energies of double helix formation are, in essence, determined by base pair stacking, and hydrogen bonding contributes only very little if at all. Specific template action is no privilege of hydrogen-bonded base pairs between nucleotides; it is found also in protein-protein interactions of the leucine zipper and has recently been used in the design of oligopeptides which are suitable for autocatalytic synthesis previously known only to occur with oligonucleotides [3]. Recent studies on natural ligand- and drug-receptor complexes [4] revealed that the importance of hydrophobic interactions has been underestimated so far, and the poor predictive power of rational design might well be a result of too much weight given to electrostatic forces, in particular to hydrogen bonding.

In essence, hydrogen bonds (X–H$\cdots Y$) differ from other intermolecular interactions in one aspect that makes them unique: only hydrogen atoms have no inner shell electrons, thus allowing for larger changes in electron densities than in other cases. The approach of a polar group towards a polar X–H bond causes strong polarization effects that give rise to well known regularities in molecular structures and spectra which are commonly used as diagnostic tools for the detection of hydrogen bonds. Hydrogen bonds, in general, are not stronger than other inter-molecular forces between polar groups or molecules, but their directive power is more pronounced. In other words, the strength of the interaction depends strongly on the relative orientations of the bond X–H and the lone pair at the atom Y because of the local nature of the dipoles involved in a hydrogen bond. The preferred geometrical arrangement of the three atoms forming the hydrogen bond, *i.e.* X, H, and Y, is linear. The directionality of hydrogen bonds provides their major contribution to biopolymer structures: the (almost) linear hydrogen bond geometry is the basis for the structures of α- and other helices, β-pleated sheets, and nucleic acid base pairs.

During the last decade several books have been published on hydrogen bonding [5–8] as well as on hydrogen bonds within the broader aspects of molecular inter-actions [9] and molecular clusters [10]. In this overview we summarize current data on hydrogen bonding and try to describe the actual picture of structure formation and stability in biopolymers. Section 2 deals with the highly accurate results obtained by spectroscopy and *ab initio* calculations. In section 3 we shall briefly report on the area of hydrogen-bonded networks in molecular clusters, a field which is very actively studied at present. The following three sections contain a state of the art review on the knowledge of biopolymer structures, stability, and formation as well as an outline of the current understanding of supramolecular complexes in biology. We end by giving an outlook to hydrogen bond research in biology.

2. Small Hydrogen-Bonded Complexes

The basic features of small hydrogen-bonded complexes were known in essence already in the seventies [11]. During the last decade the only remarkable progress was achieved in accuracy in both computational and experimental studies. Progress in computations was caused by the enormous increase in capacities and speed of the hardware that is now available at relatively low costs. Previously developed methods in the computation of electron correlation, like the *Møller-Plesset* perturbation theory and coupled cluster techniques, have become standard now in calculations of small intermolecular complexes. Basis sets that are consistent in accuracy with the correlation techniques are in common use [12]. In the water monomer, for example, the differences between calculated and experimental bond length, bond angle, harmonic vibrational frequencies, and dipole moment are less than 2%. The most successful experimental techniques are molecular-beam based electric resonance and infrared laser spectroscopy [13], which provide previously inaccessible insights into structures and dynamics of small complexes in the vapor phase.

In order to provide a representative example for the current state of the art we present a comparison of recent data for the water dimer (Fig. 1) in the vapor phase

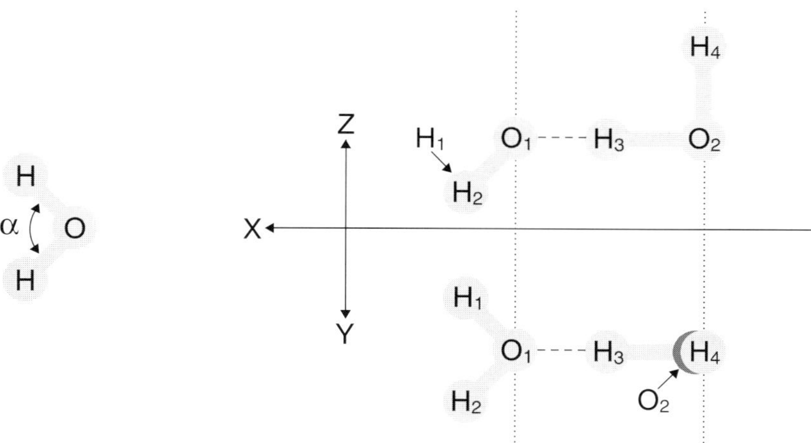

Fig. 1. Molecular geometry of the water dimer. The horizontal and vertical projection of the dimer is shown in C_s-geometry (which is frequently assumed in *ab initio* calculations); the (x,z)-plane is the plane of symmetry, $H_1H_2O_1$ is the acceptor, and $H_3H_4O_2$ the (hydrogen) donor molecule. Relative orientations of the two water molecules are described by rotations around an x-, y-, and z-axis through each of the two oxygen atoms. The *Euler*ian angles, χ, θ, and ϕ, are defined as angles between the C_2-axis of water and the corresonding coordinate axis, y, z or x, respectively. The definition of the angle α in the free water molecule is shown at the left

Table 1. The hydrogen bond in the water dimer [14]. The geometry of the complex is show in Fig. 1; values in parentheses refer to the isolated water monomer; the subscripts 0 and e refer to the vibrational ground state and the energy minimum of the dimer; 1 denotes the hydrogen acceptor and 2 the hydrogen donor molecule, and θ, ϕ, and χ are the *Euler*ian angles, respectively.

Quantity	Calculated value	Experimental value
$R_0(O_1\text{--}O_2)/\text{Å}$	–	2.976
$R_e(O_1\text{--}O_2)/\text{Å}$	2.898	2.946
$R_e(O_1\text{--}H_1)/\text{Å}$	0.960 (0.959)	–(0.957)
$R_e(O_2\text{--}H_3)/\text{Å}$	0.965 (0.959)	–(0.957)
$R_e(O_2\text{--}H_4)/\text{Å}$	0.958 (0.959)	–(0.957)
$\alpha(H_1\text{--}O_1\text{--}H_2)/°$	104.7 (104.3)	–(104.5)
$\alpha(H_3\text{--}O_2\text{--}H_4)/°$	104.4 (104.3)	–(104.5)
$\theta_1/°$	49.0	57 ± 10
$\theta_2/°$	-52.2	-51 ± 10
$\phi_1/°$	–	22 ± 8
$\chi_1/°$	–	6 ± 20
$\chi_2/°$	–	<30
$D_e/\text{kJ}\cdot\text{mol}^{-1}$	-23.32	-22.6 ± 2.9
$D_0/\text{kJ}\cdot\text{mol}^{-1}$	-14.38	-15.1 ± 2.9

[14, 15] in Table 1. In addition to the molecular geometry and the dissociation energy of the complex, vibration frequencies were also calculated. Both the equilibrium dissociation energy and the zero point contribution from vibrations agree well with the experimental data: errors are in the range of one kJ/mol or less. The shift in the harmonic bond stretching frequency caused by the hydrogen bond is computed to be $-171\,\text{cm}^{-1}$ and is almost identical with data from matrix isolation ($-169\,\text{cm}^{-1}$). Despite the already respectable accuracy achieved, the race for still higher precision is going on: new computational studies are dealing with improved accuracy in calculations of electronic properties [16, 17] and compete with results from near-infrared laser spectroscopy [18]. Recent studies on nuclear motion [19] gave evidence, nevertheless, on insufficient accuracy of energy surfaces for the excited states. The full spectroscopic information on the water dimer has also been used to derive a pair potential function that takes into account polarization of the water molecule [20].

A considerable number of other hydrogen-bonded homo- and heterodimers has been studied in the vapor phase, some of them with an accuracy comparable to that achieved for the water dimer. As examples we mention recent investigations of methanol clusters by infrared cavity ringdown spectroscopy [21] and a review including also clusters of hydrazine [22].

3. Nonadditivity, Clusters, and Hydrogen-Bonded Networks

Two classes of cooperative phenomena resulting in deviations from additivity of free energies are observed with hydrogen bonds coupled to polarizable electron systems: (*i*) resonance assisted hydrogen bonding and (*ii*) chains, cycles, or other networks of hydrogen bonds. In both cases hydrogen bonds become stronger as a

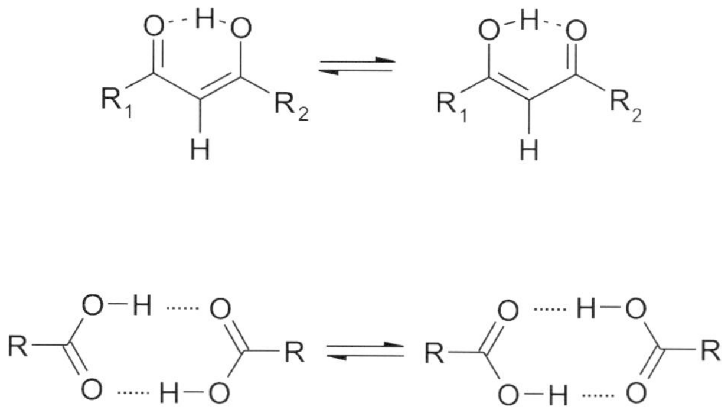

Fig. 2. Conjugation assisted hydrogen bonds. Two examples are shown: (*i*) the intramolecular hydrogen bond in the enol-form of an 1,3-dicarbonyl compound, and (*ii*) the two intermolecular hydrogen bonds in a carboxylic acid dimers. In both cases, the hydrogen bond is reinforced and proton transfer is facilitated by π-electron conjugation

result of coupling to the molecular environment. In other words, mutual polarization increases the strength of the hydrogen bond.

The first case is typical for intermolecular hydrogen bonds or hydrogen bonds in dimers where donor and acceptor are connected by one or more bonds with mobile, *i.e.* easily polarizable, electrons. These are commonly one or more conjugated π-electron bonds. The reinforcement of hydrogen bonds through cyclic conjugation is often characterized as resonance assisted hydrogen bonding. Representative examples are the hydrogen bonds in the enol-forms of 1,3-dicarbonyl compounds or the cyclic dimers of carboxylic acids (Fig. 2). Some cases are discussed in the contributions by *Koll* and *Wolschann* as well as by *Wolf et al.* in this issue (see also Ref. [23]). Other examples for resonance of coupled hydrogen bonds are the base pairs of nucleic acids in various geometrical arrangements [5]. They will also be mentioned in sections 4 and 5. Proton transfer along cycles of hydrogen bonds is facilitated when it is coupled to long-range displacements in electron distributions. Examples are proton transfer in the frequently studied dimers of carboxylic acids, water mediated proton transfer in formamide [24], and proton transfer in malonaldehyde [23].

The structures of molecular clusters provide not only information on the nature of the forces which shape these stable complexes of several molecules; they are also the most important tools for understanding and modelling of pure liquids and solutions. Both experimentalists and theorists have treated molecular clusters as a bridge between the gas phase and condensed matter. A survey consisting of a series of reviews on various clusters has been published recently [10]. Clusters in which hydrogen bonding determines structures and intermolecular energies of the aggregates are of particular interest since hydrogen-bonded liquids and protic solutions show features that are completely missing in *van der Waals* clusters or aprotic solvents. The behaviour of hydrogen-bonded clusters is dominated by the features of the hydrogen bond and reflects therefore also the influence of the involved molecules on the hydrogen bond strength and the related energetic properties. The importance of these hydrogen-bonded systems is clearly documented by the huge amount of investigations and publications on this topic. Some extended reviews appeared just recently, collecting the experimental and theoretical studies of the last years [5, 9, 25–28]. These studies focus on the structures and the thermodynamic properties of the association complexes in order to provide a basis for understanding cooperativity in hydrogen-bonded networks, structural rearrangements, and proton transfer dynamics. Cooperative phenomena in hydrogen-bonded chains and networks result from mutual polarization between hydrogen bonds. Unusual strength of polarization and easy displacement of protons are consequences of the already mentioned peculiarity of the hydrogen atom resulting from the absence of inner shell electrons. Larger water clusters and their vibrational spectra have been predicted by *ab initio* calculations [14, 29] and compared to experiments [30–34]. A significant contraction of the OO-distance with increasing cluster size has been observed. The dynamics of linear water chains has been studied recently by *ab initio* molecular dynamics [35]. Diffusion dynamics and tunnelling in vibration-rotation spectra have also been considered recently [27, 28, 36–38].

Cyclic conformations of hydrogen-bonded chains are of particular interest since they show increased hydrogen bond strength and are assumed to exist in liquid water [14, 29]. Changes in the geometry of hydrogen-bonded networks are often considered in explanations for the peculiar properties of water like the increase of density between $0°$ and $4°C$ or the high mobility of protons and hydroxyl ions. Large clusters of up to 35 water molecules have been studied by less accurate computational methods [39]. These investigations try to find answers to questions how clustering influences the properties of individual molecules and whether or not the obtained results are suitable for extrapolations to large or infinite ensembles.

Studies on methanol clusters gave results similar to those obtained for water clusters: aggregates up to the size of four molecules exist in mainly cyclic conformation. This was shown by *ab initio* as well as by density functional methods [40–42] and experimentally verified by infrared cavity ringdown spectroscopy [21] as well as by vibrational spectroscopy [43, 44]. Pressure induced strengthening and temperature dependent weakening of hydrogen bonding in methanol clusters has been investigated by NMR spectroscopy [45]. Many extensive studies report on various association complexes of water and methanol molecules with other compounds from which we cite only a few examples: water and methanol with neutral molecules [46, 47], water with protons and other cations [47–49], and water with anions [12]. Finally and as a kind of curiosity, we mention a recent report on the formation of linear HCN chains in superfluid helium [50].

4. Hydrogen Bonds and Biopolymer Structures

The role of hydrogen bonding in biological structures was treated comprehensively in the monograph by *Jeffrey* and *Saenger* [5]. Here, we mention only some facts on the balance of hydrogen bonding within biopolymers and solvation of the molecules by water. Recent developments in the construction of supramolecular complexes provide new insights into the old problem of enthalpy-entropy compensation phenomena in aqeous solutions [51, 52].

The directive power of hydrogen bonds is apparently the major factor for the uniqueness and specificity of biopolymer structures. This commonly accepted fact creates a puzzle: biopolymers form their specific native structures only in aqueous environments, but hydrogen bonding between acceptor and donor molecules is usually ineffective in aqueous solutions because of the excellent donor and acceptor properties of the water molecule [53]. Free energies of interaction reflect the differences in hydrogen bonding to the binding partner and the solvent, and they are commonly very small in aqueous solutions. In order to overcome this problem, chemists are constructing locally hydrophobic environments in order to exclude mobile water molecules [54, 55]. The hydrogen bond strength then becomes fully available, and highly specific and remarkably complex structures can be constructed. Obviously, hydrogen bonding and space filling are the two principles for the assembly of stable aggregations [56]. Another trick to improve hydrogen bonding is orientation of molecules on surfaces [53].

The same two principles, hydrogen bonding and space filling, are sufficient to explain the structures of biomolecules. The driving force for space filling is hydrophobic interaction that can be interpreted as a consequence of hydrogen bonding in the surrounding water. Proteins and nucleic acids form compact structures in water because of their tendency to minimize the area of the boundary to the solvent. Although there is no doubt concerning the nature of the forces stabilizing biopolymer conformations, the relative weights of the contributions are not yet known with sufficient accuracy. Hydrogen bonds define the specific geometry of secondary structures in proteins, in particular, in the α-helix and in other hydrogen-bonded helical conformations[1] or in the two different β-sheet conformations. Helix formation in polypeptides and proteins, on the other hand, is not dependent on hydrogen bonding. The two different polyproline helices (I and II) contain no hydrogen bonds since the nitrogen atom in the proline residue carries no hydrogen. Other examples are the various coiled coil structures built from α-helices. They are stabilized exclusively by the constraint of optimal packing as a result of hydrophobic forces.

Double helix formation in nucleic acids (*DNA* and *RNA*) is driven by base pair stacking and not by hydrogen bond formation. As mentioned before, the free energy contribution of hydrogen bonds to the base pairing energies in aqueous solutions is given by the difference in free energies between the hydrogen bonds in the complex and the hydrogen bonds to the solvent in the isolated molecules. Experiments performed in the seventies on double helix formation within one or between two *RNA* molecules [57] have shown indeed that the major contribution to the stability of double helical regions comes from base pair stacking rather than from hydrogen bond formation. These reactions can be described well by cooperative stack formation thermodynamics as expressed by Eq. (1) where K is the macroscopic equilibrium constant for stack formation, s is the microscopic constant for the conversion of a coil element into a segment of the double helix, σ is the nucleation parameter, and n is the length of the stack.

$$K(n) = \sigma \cdot s^n \qquad (1)$$

Approximate values at $T = 0°C$ are $\sigma = 10^{-3}$, $s = 10$ for an **AU** and $s = 100$ for a **GC** pair. Accordingly, the formation of the first stable nucleus ($K(n) > 1$) requires four **AU** or two **GC** pairs, respectively.

---

[1] The best known examples are the 3.0_{10}- and the π-helix which, respectively, contain one amino acid less or one amino acid more than the α-helix between the >CO and the >NH group in the hydrogen-bonded loop.

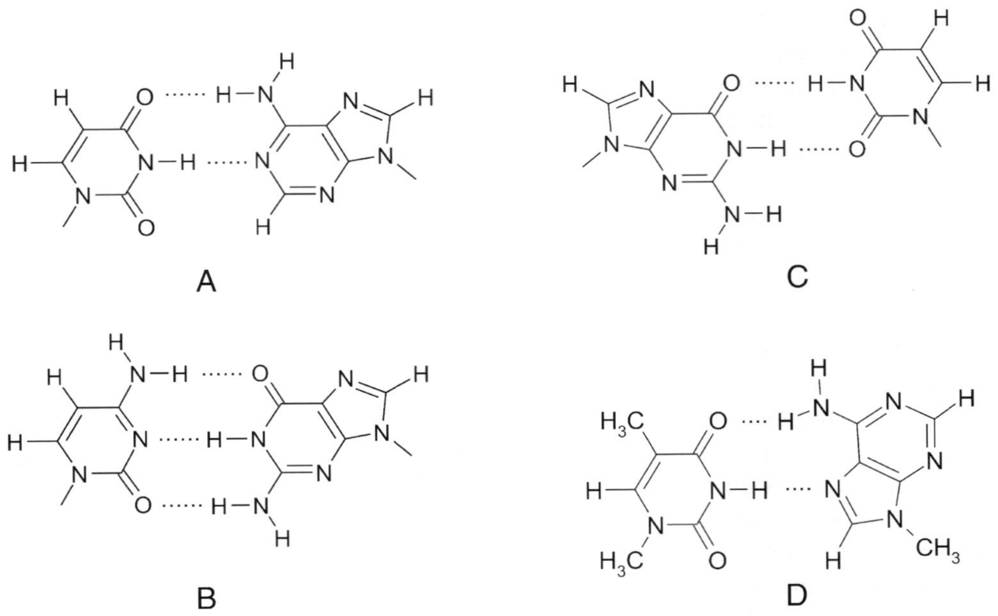

A

B

C

D

Fig. 3. Base pairing patterns of nucleotides. A and B represent the two *Watson-Crick* base pairs, **AU** and **GC**, respectively; the wobble pair **GU**, which occurs regulary in the A- and A'-conformation of *RNA*, is shown in C. D sketches a *Hoogsteen* pair between 1-methylthymine and 9-methyladenine. Gray cycles indicate the positions at which the base pairs are connected to the ribose-phosphate backbone

Base pair stacking – unlike conventional hydrophobic interaction – is an enthalpy driven process. The driving force for stacking, however, is by no means less sophisticated than that for hydrophobic aggregation. The detailed molecular mechanism involves water structure in both cases: base or base pair stacking is not observed in non-aqueous media like, for example, chloroform. Stacking kinetics of nucleotide bases in absence of a ribose phosphate backbone has been studied by means of ultrasound absorption. These studies on N_2, N_9-dimethyladenine provide precise information on the thermodynamic parameters of nucleotide base stacking [58]. First encountered in the crystal structures of transfer-*RNA*s, end-to-end or coaxial stacking of double helical regions has been observed in most other structures determined so far, too [59, 60]. Extension of double helices in order to minimize contact areas between water and nucleotide bases seems to represent a general and highly relevant principle of *RNA* structure formation.

In addition to the long known conformations of base pairs, *i.e. Watson-Crick*, *Hoogsteen*, and wobble-**GU** (Fig. 3), a whole series of other hydrogen-bonded structures were found in the crystals and NMR spectra of *RNA* molecules. Double helices are often extended through a pair of purine bases (**AA**, **GA**, or **GG**) [60, 61]. Another hydrogen-bonded interaction, observed mainly in large interior loops, is the **UU** base pair [62–64]. These pyrimidine-pyrimidine pairs occur also in small clusters, *e.g.* (**UU**)$_n$ with $n = 2$, 3. Base interactions are not confined to pairs; base triples (Fig. 4) occur regularly, and several examples are already known from crystal structures of transfer-*RNA*s. Quartets of **G**s and **A** platforms [60] represent

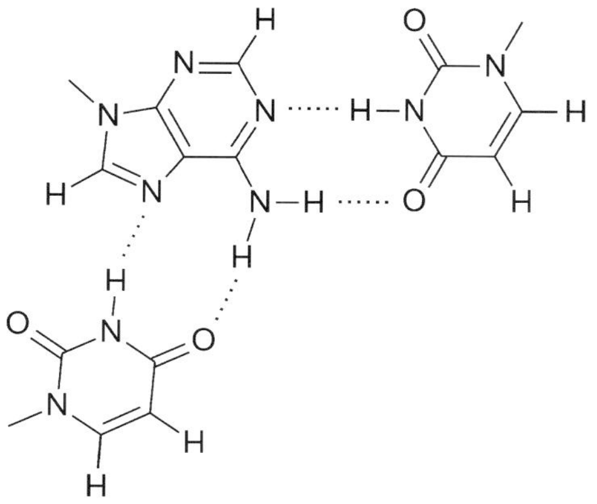

Fig. 4. Conformation of the uracil-adenine-uracil trimer. The trimer represents a combination of a *Watson-Crick* and a *Hoogsteen* base pair. It can be readily incorporated into a triple helix as it is found in the poly-**A**.(poly-**U**)$_2$ aggregate

examples of interactions between four bases with regular structures. In this context we mention also an extensive *ab initio* study on hydrogen bonding and stacking of nucleic acid bases and base pairs which was undertaken in order to derive potentials of mean force for nucleotide bases [65]. Contributions of the aqueous medium were taken into account by means of a *Langevin* dipole solvation model.

Template action is the basis of nucleic acid replication and thus fundamental to biology. *DNA* and *RNA* replication is based on the complementarity of the bases in *Watson-Crick* base pairs (Fig. 3). Hence, template action is determined by hydrogen bonding, and it was indeed possible to synthesize self-complementary oligonucleotides through the autocatalytic reaction shown in Fig. 5 without the help of enzymes [66]. All other chemical systems showing template action made use of hydrogen bonding in suitable molecular environments. Recently, template action was extended to oligopeptides [3]. In this example, complementarity is not based on patterns of hydrogen bonds but on space filling (Fig. 5). For two α-helices with complementary patterns of leucine and valine residues on their hydrophobic sides an autocatalytic replication process based on this "knobs into holes" – interaction has been designed and successfully realized.

The exploration of the structures of supramolecular complexes is one of the most remarkable successes of recent structural biology. Structures of cellular particles as complex as the ribosome will soon be available at atomic resolution [67]. The structures of several virus particles or so-called virions have been determined at sufficiently high resolution for the identification of their molecular geometry [68].

5. Folding of Biopolymers

The conventional view of protein folding was initiated by the results of experiments on the denaturation of small proteins performed in the sixties and early seventies [69, 70]. Several small protein molecules like ribonuclease S undergo reversible denaturation with a simple two step kinetics, $N \rightleftharpoons C$, showing

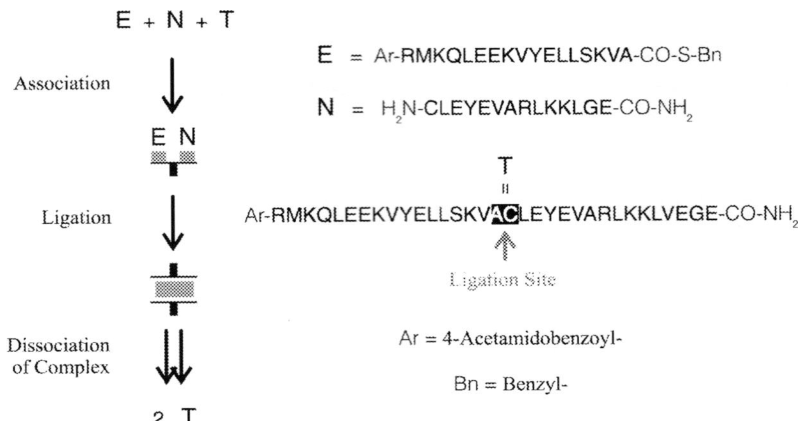

Autocatalytic Replication of an Oligopeptide

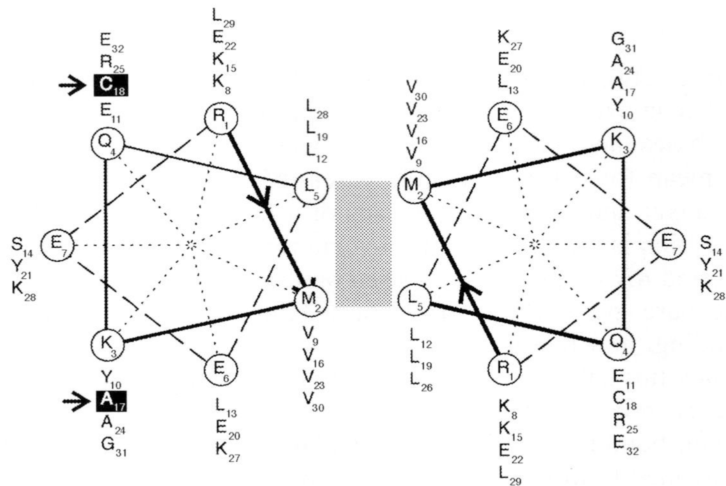

Helical-Wheel Diagram of Template-Ligand Interactions

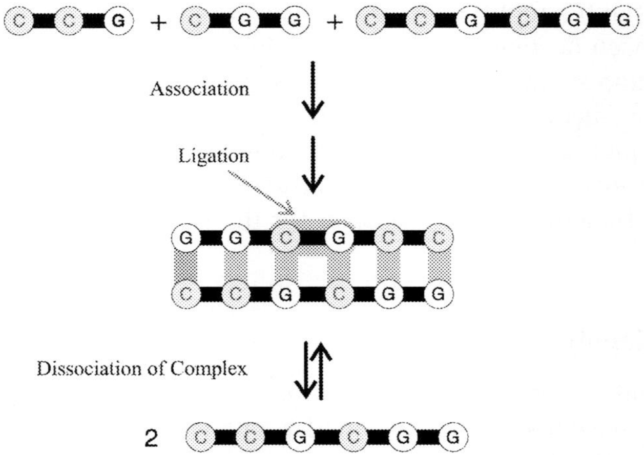

Autocatalytic Replication of a Hexanucleotide

that only the native state (*N*) and the random coil state (*C*) are significantly populated. Experimental studies on protein-ligand binding at low temperatures [71] revealed, however, that not only the populations in loose random coils but also in the rigid native states of proteins are highly heterogeneous. What has been

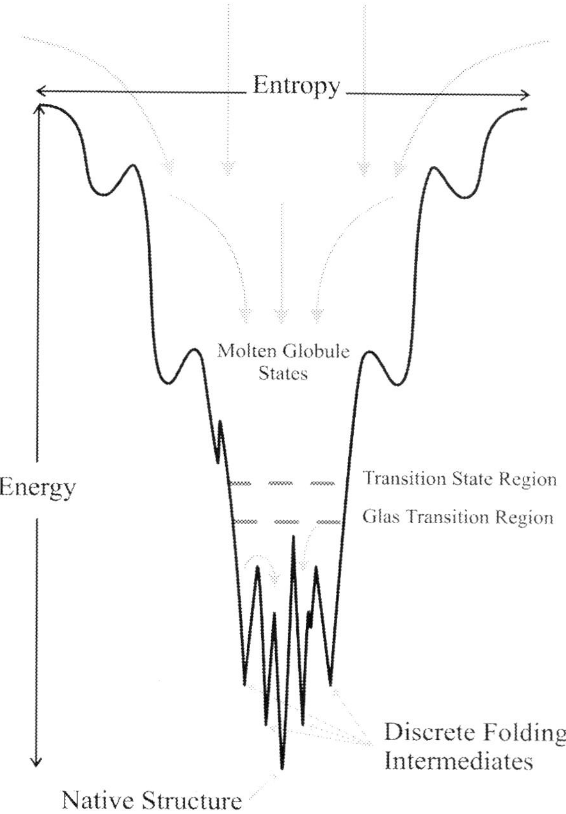

Fig. 6. Funneling trajectories of protein folding. Proteins have extremely large numbers of conformations and hence random coils would not be able to fold into the native states within reasonable times unless there are folding pathways guiding the molecules towards the energy minimum [72]. A statistical theory of protein conformations provides an explanation for the folding problem [73]: a very large number of initial random coil conformations gives rise to a smaller number of partially folded intermediate states which in turn form a still smaller number of compact states. The diversity of pathways leading towards the native state is reduced with decreasing energy, and eventually the molecule ends up in the native conformation

Fig. 5. Templates in molecular replication. Two examples of autocatalytic oligomer formation are shown. The upper part presents a reaction on an oligopeptide template where the aggregation is caused by hydrophobic forces [3]. Specificity of the interaction results from "knobs into holes" space filling of valine and leucine residues which appear on opposite sides on the two α-helices forming the coiled coil (see the **V** and **L** residues in the helical-wheel diagram). The lower part of the figure shows the template induced autocatalytic synthesis of the hexamer **CCGCGG** [66] from trimers. Template interaction of oligonucleotides results from conventional *Watson-Crick* type hydrogen bonding between **G** and **C**

considered as a single distinct conformation turned out as a family of structurally similar conformations, each corresponding to an individual local minimum of the conformational energy landscape. Conformational diversity is reduced during the folding process that eventually ends in the native state [72, 73]. This idea of a very large number of folding pathways merging into fewer and fewer trajectories during the approach to the thermodynamically most stable state has been characterized as the folding funnel paradigm (Fig. 6).

Folding of protein molecules starts through nucleation of secondary structure elements, commonly α-helices, and leads to an intermediate form that is already condensed but still shows a certain degree of mobility [74]. This intermediate has been characterized as molten globule and contains a high fraction of the hydrogen bonds found in the native structure. In case of single stranded *RNA* molecules folding is initiated by the formation of double helical regions or stacks making up the so-called secondary structure which is rigidified by the formation of tertiary contacts and presumably represents a folding intermediate as the molten globule does for protein folding.

6. Conclusions and Outlook

Investigations and computational studies on small molecular systems have reached a previously unimaginable accuracy. This holds in particular also for small intermolecular complexes including those involving hydrogen bonds. Errors as small as a few percent in molecular structures, energies, and vibrational frequencies have been achieved in recent calculations. The quality of computed data competes successfully with the results from experimental studies which are mainly based on molecular beam electric resonance and IR laser spectroscopy. For larger systems the achievable accuracy is less, but computations are still useful. Investigations on hydrogen-bonded clusters of many molecules are currently in progress and reveal more and more accurate data on large aggregates that show almost bulk-like properties in the interior. In the future these studies will eventually lead to a better understanding of associated liquids and protic solutions.

Despite impressive successes in the prediction of biopolymer structures the theory still suffers from insufficient or even inadequate treatment of the solvent. In molecular dynamics simulations of biopolymers a few thousand water molecules are added, but often this is even not enough for full coverage of the molecular surface. Further progress in the theory of hydrogen-bonded liquids, in particular water, will be of great help in the theory of biopolymer structures. The hydrophobic force which leads to condensation of proteins in water is a consequence of the peculiar properties of liquid water. The same is true for base pair stacking that provides the major force in the formation of nucleic acid structures.

Protein structures are the result of a delicate balance between hydration of polar groups at the surface and hydrophobic collapse of the residues buried in the interior. The current theory of protein folding is based on the statistical mechanics of this collapse. It allows to analyze and to understand the role of hydrogen bonding in the nucleation complexes as well as in later condensed states like the molten globule. *RNA* folding is less well understood than protein folding but shows striking similarities: the conventional hydrophobic force is replaced by the stacking interaction of nucleotide bases and base pairs. Loose *RNA* conformations with fully developed secondary structures are, presumably, the counterparts of molten globules.

References

[1] Pauling L, Corey RB, Branson HR (1951) Proc Natl Acad Sci USA **37**: 205; Pauling L, Corey RB (1951) Proc Natl Acad Sci USA **37**: 251 and 729

[2] Watson J, Crick FHC (1953) Nature **171**: 737

[3] Lee DH, Granja JR, Martinez JA, Severin K, Ghadiri R (1996) Nature **382**: 525

[4] Davis AM, Teague SJ (1999) Angew Chem Int Ed **38**: 736

[5] Jeffrey GA, Saenger W (1991) Hydrogen Bonding in Biological Structures. Springer, Berlin

[6] Smith DA (ed) (1994) Modeling the Hydrogen Bond, vol 569. American Chemical Society, Washington, DC

[7] Scheiner S (1997) Hydrogen Bonding. A Theoretical Perspective. Oxford University Press, NY

[8] Jeffrey GA (1997) An Introduction to Hydrogen Bonding. Oxford University Press, NY

[9] Scheiner S (ed) (1997) Molecular Interactions. From van der Waals to Strongly Bond Complexes. Wiley, Chichester, UK

[10] Bowman JM, Bačić Z (eds) (1998) Molecular Clusters. Advances in Molecular Vibrations and Collision Dynamics, vol 3. JAI Press, Stamford, CT

[11] Schuster P, Zundel G, Sándorfy C (1976) The Hydrogen Bond – Recent Developments in Theory and Experiment, vol I–III. North Holland, Amsterdam

[12] Xantheas SS, Dunning TH Jr (1998) *Ab Initio* Characterization of Water and Anion-Water Clusters. In: Ref. [10], p. 281

[13] Weber A (ed) (1987) Structure and Dynamics of Weakly Bonded Molecular Complexes. Reidel, Dordrecht

[14] Xantheas SS, Dunning TH Jr (1993) J Chem Phys **99**: 8774

[15] Mack KM, Muenter JS (1977) J Chem Phys **66**: 498

[16] Tschumper GS, Kelty MD, Schaefer III HF (1999) Mol Phys **96**: 493

[17] Klopper W, Lüthi HP (1999) Mol Phys **96**: 559

[18] Huang ZS, Miller RE (1997) J Chem Phys **91**: 6613

[19] Leforestier C, Braly LB, Liu K, Elrod MJ, Saykally RJ (1997) J Chem Phys **106**: 8527

[20] Fellers RS, Leforestier C, Braly LB, Brown MG, Saykally RJ (1999) Science **284**: 945

[21] Provencal RA, Paul JB, Roth K, Chapo C, Casaes RN, Saykally JR, Tschumper GS, Schaefer III HF (1999) J Chem Phys. **110**: 4258

[22] Buck U (1998) Vibrational Spectroscopy of Small Size-Selected Clusters. In: Ref. [10], pp 127–161

[23] Wolf K, Mikenda W, Nusterer E, Schwarz K, Ulbricht C (1998) Chem Eur J **4**: 1418

[24] Kim Y, Lim S, Kim H-J, Kim Y (1999) J Chem Phys A **103**: 617

[25] Deakyne CA (1997) Conventional and Unconventional Hydrogen-Bonded Ionic Clusters. In: Ref. [9], p 217

[26] Karpfen A (1997) Case Studies in Cooperativity in Hydrogen-Bonded Clusters and Polymers. In: Ref. [9], p 265

[27] Gregory JK, Clary DC (1998) Diffusion Monte Carlo Studies of Water Clusters. In: Ref. [10], pp 311–364

[28] Wales DJ (1998) Rearrangements and Tunneling in Water Clusters. In: Ref. [10], p 397

[29] Klopper W, Schütz M (1995) Ber Bunsenges Phys Chem **99**: 469

[30] Dykstra CE (1999) Chem Phys Letters **299**: 132

[31] Graf S, Leutwyler S (1998) J Chem Phys **109**: 5393

[32] Jongseob Kim, Kwang S Kim (1998) J Chem Phys **109**: 5886

[33] Hartke B, Schütz M, Werner HJ (1998) Chem Phys **239**: 561
[34] Kryachko E (1998) Int J Quant Chem **70**: 831
[35] Mei HS, Tuckerman ME, Sagnella DE, Klein ML (1998) J Phys Chem B **102**: 10446
[36] Sabo D, Bacic Z, Graf S, Leutwyler S (1998) J Chem Phys **109**: 5404
[37] Brown MG, Keutsch FN, Saykally RJ (1998) J Chem Phys **109**: 9645
[38] Liem X Dang (1999) J Chem Phys **110**: 1526
[39] Khan A (1999) J Phys Chem A **103**: 1260
[40] Buck U, Ettischer I (1998) J Chem Phys **108**: 33
[41] Buck U, Siebers JG, Wheatley RJ (1998) J Chem Phys **108**: 20
[42] Mo O, Yanez M, Elguero J (1997) J Chem Phys **107**: 3592
[43] Hagemeister FC, Gruenloh CJ, Zwier TS (1998) J Phys Chem A **102**: 82
[44] Parra RD, Zeng XC (1999) J Chem Phys **110**: 6329
[45] Czeslik C, Jonas J (1999) Chem Phys Letters **302**: 633
[46] Zwier TS (1998) The Infrared Spectroscopy of Hydrogen-Bonded Clusters: Chains, Cycles, Cubes, and Three-Dimensional Networks. In: Ref. [10], p 249
[47] Ebata T, Fujii A, Mikami N (1998) Int Rev Phys Chem **17**: 331
[48] Marx D, Tuckerman ME, Hutter J, Parrinello M (1999) Nature **397**: 601
[49] Brugé F, Bernasconi M, Parinello M (1999) J Chem Phys **110**: 4734
[50] Nauta K, Miller RE (1999) Science **283**: 1895
[51] Lumry R, Rajender S (1970) Biopolymers **9**: 1125
[52] Lauffer MA (1975) Entropy-Driven Processes in Biology. Polymerization of Tobacco Mosaic Virus Protein and Similar Reactions. Springer, Berlin
[53] Ariga K, Kunitake T (1998) Acc Chem Res **31**: 371
[54] Rotello VM, Viana EA, Deslongchamps G, Murray BA, Rebek J Jr (1993) J Am Chem Soc **115**: 797
[55] Torneiro M, Still WC (1995) J Am Chem Soc **117**: 5887
[56] Martin T, Obst U, Rebek J Jr (1998) Science **281**: 1842
[57] Pörschke D (1977) Elementary Steps of Base Recognition and Helix-Coil Transitions in Nucleic Acids. In: Pecht I, Rigler R (eds) Chemical Relaxation in Molecular Biology. Springer, Berlin
[58] Pörschke D, Eggers F (1972) Eur J Biochem **26**: 490
[59] Pley HW, Flaherty KM, McKay D (1994) Nature **372**: 68
[60] Cate JH, Gooding AR, Podell E, Zhou K, Golden BL, Kundrot CE, Cech TR, Doudna JA (1996) Science **273**: 1678
[61] Huang S, Wang Y-X, Draper DE (1996) J Mol Biol **258**: 308
[62] Baeyens KJ, De Bondt HL, Holbrook (1995) Nat Struct Biol **2**: 56
[63] Wang Y-X, Huang S, Draper DE (1996) Nucleic Acids Res **24**: 2666
[64] Lietzke SE, Barnes CL, Berglund JA, Kundrot CE (1996) Structure **4**: 917
[65] Florián J, Šponer J, Warshel A (1999) J Phys Chem B **103**: 884
[66] von Kiedrowski G (1986) Angew Chem Int Ed Engl **25**: 932
[67] Green R, Noller HF (1997) Ann Rev Biochem **66**: 679
[68] Wah C, Burnett RM, Garcea RL (1997) Structural Biology of Viruses. Oxford Univ Press, NY
[69] Tanford C (1961) Physical Chemistry of Macromolecules. Wiley, New York
[70] Anfinsen CB (1973) Science **181**: 223
[71] Frauenfelder H, Parak F, Young RD (1988) Ann Rev Biophys Biophys Chem **17**: 451
[72] Dill KA, Chan HS (1997) Nat Struct Biol **4**: 10
[73] Bryngelson JD, Onuchic JN, Socci ND, Wolynes PG (1995) Proteins Struct Funct Gen **21**: 167
[74] Freire E (1995) Ann Rev Biophys Biomol Struct **24**: 141

Received May 3, 1999. Accepted May 4, 1999

NMR Techniques to Study Hydrogen Bonding in Aqueous Solution[a]

Robert Konrat[*]**, Martin Tollinger, Georg Kontaxis**[b]**, and Bernhard Kräutler**

Institute of Organic Chemistry, University of Innsbruck, Austria

Summary. Recent improvements in NMR methodology have significantly increased the scope of hydrogen bond related problems that can be now addressed by solution NMR methods. A growing number of applications are exploiting these NMR techniques to study complex molecular systems and elicit otherwise inaccessible information on hydrogen bonding in aqueous solution.

Keywords: Hydrogen bond; NMR Spectroscopy; Chemical shift anisotropy; Isotopic fractionation factor; Hydrogen exchange.

NMR-Spektroskopie von Wasserstoffbrücken in wäßriger Lösung

Zusammenfassung. Kürzlich erarbeitete methodische Weiterentwicklungen der Kernresonanzspektroskopie erlauben nunmehr auch Untersuchungen von Wasserstoffbrücken in wäßriger Lösung. Diese neuartigen experimentellen Methoden wurden bereits erfolgreich angewandt, um geometrische und energetische Wasserstoffbrückenparameter in Lösung zu bestimmen.

Introduction

The hydrogen bond [1–5] is considered to be not only a significant determinant of the solution structure of biomolecules, but also of crucial importance to all biochemical processes by contributing to transition state stabilization or to ligand binding specificity. Hydrogen bonds also control chemical self-assembly processes by which 'programmed' molecular subunits spontaneously form complex supramolecular frameworks [6, 7]. The hydrogen bond A-H$\cdots B$ is typically described as an electrostatic attraction between the positive end of the bond dipole of A-H and a centre of negative charge on B [4]. In a typical situation, A is sufficiently electronegative to ensure a strongly polar bond. The acceptor site on B is typically

---

[*] Corresponding author
[a] Dedicated to Prof. *H. Gruber* on the occasion of his 70th birthday
[b] Present address: Laboratory of Chemical Physics, National Institute of Diabetes and Digestive and Kidney Diseases, National Institute of Health, Bethesda, Maryland, USA

defined as a lone pair. Thus, the hydrogen bond inherently involves the sharing of hydrogen atoms to varying extents with other atoms, typically leading to a lengthening of the donor H-bond and shortening overall the distance to acceptor B. It is generally assumed that the hydrogen bond is a weak interatomic interaction, and that it is an order of magnitude weaker than covalent bonding. However, there are cases (*e.g.* charged species in the gas phase) where the strength of hydrogen bonds are comparable to those of covalent bonds. Ionization greatly affects hydrogen bonding. If the donor is positively charged $(A\text{-}H)^+$, there will be an increased electrostatic attraction between the donor and the acceptor B. By the same token, a negatively charged acceptor B will strengthen the hydrogen bonding.

Bond lengths, energetics, vibrational frequencies, electron distributions, and various other spectroscopic characteristics are modified upon formation of a hydrogen bond. The structural characteristics of hydrogen bonds (*e.g.* hydrogen bond lengths, proton location, bond angles) are studied by X-ray or neutron diffraction techniques. Based on X-ray studies, a correlation was established between the distances $R_{A\text{-}B}$ and $R_{A\text{-}H}$. The most notable observation is the decrease in $R_{A\text{-}B}$ accompanied by an increase in $R_{A\text{-}H}$ [3, 4]. This relationship also serves to classify hydrogen bonds. Typically, weak bonds are characterized by $R_{A\text{-}B} > 260\,\text{pm}$, the hydrogen essentially staying close to its directly attached atom [3]. Strong hydrogen bonds show $R_{A\text{-}B}$ values in the range of 245–260 pm. In this case, the proton tends to move from the donor atom A towards the acceptor atom B [3]. Finally, very strong hydrogen bonds have $R_{A\text{-}B} < 245\,\text{pm}$, and the proton is located roughly halfway between the two heteroatoms [3].

Experimental information on hydrogen bond energies are available from measurements either in the gas phase [3, 8, 9] or under nonaqueous conditions [10–13]. From measurements by ion cyclotron resonance spectroscopy [3], the energies of a broad range of hydrogen bonds could be determined with values up to and exceeding $84\,\text{kJ} \cdot \text{mol}^{-1}$ (in charge localized and ionic species in the gas phase) [3], depending on the length of the hydrogen bond and the distance between the hydrogen bond donor and acceptor atoms. Empirically it was found that as a hydrogen bond becomes shorter it becomes stronger. For example, $(H_2O)_2$ has a $R_{O\text{-}O}$ distance of 297 pm and a hydrogen bond energy $E(OHO)$ of about $21.7\,\text{kJ} \cdot \text{mol}^{-1}$ [3]. In contrast, 3-(4′-biphenyl)pentane-2,4-dione has a $R_{O\text{-}O}$ distance of 244 pm and a $E(OHO)$ of about $117\,\text{kJ} \cdot \text{mol}^{-1}$ [3]. Electrostatic attraction accounts for the energies of weak hydrogen bonds. When a molecule forms a hydrogen bond, the vibrational modes are considerably changed and can be monitored using either IR or *Raman* spectroscopy where the stretching vibration of the donor bond, $\nu(AH)$, decreases in frequency. In certain cases, the shift can amount to hundreds of wavenumbers, and this change has been related to hydrogen bond energies (Ref. [3] and references cited therein).

In this review we will focus on and limit the discussion to NMR based techniques to characterize hydrogen bonding. Starting from already established spectroscopic manifestations such as characteristic downfield shifts of the proton resonances [3] or the isotope fractionation factor φ [3], we will discuss intermolecular exchange rate measurements (k_{ex}) of exchangeable protons and introduce solution measurements of the chemical shift anisotropy (CSA) of hydrogen as sensitive probes for hydrogen bonding in aqueous solution.

Chemical Shift and Chemical Shift Anisotropy

As a rule, hydrogen bonding leads to a downfield shift of the NMR resonance of the hydrogen bonded hydrogen [3]. Solid state NMR revealed a qualitative correlation between the hydrogen bond length and the downfield shift of the isotropic chemical shifts (σ_{iso}) of the hydrogen bonded hydrogens [3]. There is an extensive literature about correlations between the easily accessible spectral parameter σ_{iso} and other hydrogen bond related parameters, *e.g.* hydrogen bond distances, variations in the ν_{AH} stretching frequencies, linear stretching force constant, and quadrupolar coupling constant (e^2qQ/h). In liquids, the proton chemical shift can be used to calculate thermodynamic parameters associated with hydrogen bonding [3].

The two nearby nuclei in the hydrogen bond A-H$\cdots B$ cause the ^{1}H NMR signal of A-H$\cdots B$ to occur further downfield than the resonance of a similar proton A-*H* that cannot form a hydrogen bond [3]. This deshielding effect was explained by *ab initio* calculations on hydrogen bonded O-H$\cdots$O systems [14, 15]. The deshielding of the isotropic chemical shift was explained by three effects. First, the proton loses electron density upon hydrogen bond formation, and due to the fact that there are only *s* functions on the hydrogen, the chemical shift tensor components are deshielded isotropically. In addition, the acceptor oxygen deshields the perpendicular components but shields the parallel component. Third, the oxygen atom of the hydrogen bond donor accounts for some deshielding in the perpendicular components and can either shield or deshield the parallel component. In sum, this leads to a net deshielding of both the isotropic chemical shift and the perpendicular components of the chemical shift tensor. Multiple pulse NMR experiments of *e.g.* O-H$\cdots$O systems in the solid state [16] confirmed the predicted correlation between the isotropic chemical shift σ_{iso} and the O$\cdots$O separation $R_{O\cdots O}$. In this study, a maximum downfield shift of about 10 ppm for σ_{iso} was found. The σ_{iso} value was suggested to be the preferable NMR probe for hydrogen bonding, as it is least sensitive to extraneous effects (*e.g.* nearby ions or atoms in the solid state environment). However, measurements of the individual tensor components revealed a more detailed picture of the effect of hydrogen bonding geometry (*e.g.* $R_{O\cdots O}$) on spectral parameters. It was found, that the perpendicular components of the chemical shift tensor, $\sigma_\perp$, are very sensitive to $R_{O\cdots O}$, whereas the parallel component $\sigma_\parallel$ is almost unchanged. A significant correlation between $\sigma_\perp$ and the quadrupolar coupling constant e^2qQ/h was also observed, and since e^2qQ/h correlates very well to the hydrogen bond strength, $\sigma_\perp$ is largely responsible for the observed changes in σ_{iso} and in the chemical shift anisotropy (CSA) $\Delta\sigma = \sigma_\parallel - \sigma_\perp$ as the degree of hydrogen bonding changes.

Upon hydrogen bond formation involving (protein) amide protons H^N, a similar redistribution of electron density takes place resulting in a more polar charge distribution (*e.g.* favouring the more polar resonance structures). The direction of the electron density shift from the NH to the carbonyl group results in a decreased magnetic shielding for the amide proton and hence results in a shift to lower field of its ^{1}H NMR signal. This downfield shift of the isotropic amide proton signal is accompanied by an increase in chemical shift anisotropy (CSA) upon hydrogen bond formation. In peptidic systems, there exist only three solid-state NMR studies of backbone amide H^N CSA values [17–19]. All data indicate that, although the H^N CSA tensor can deviate from axial symmetry, the most shielded tensor component is aligned along the N–H bond. Recently, the magnitudes and orientations of the principal elements of the ^1H chemical shift interaction tensor of the NH side-chain protons of ^{15}N-labeled tryptophan and histidine residues were also determined [20]. Hence, ^1H chemical shift anisotropy $\Delta\sigma = \sigma_{\parallel} - \sigma_{\perp}$ should be a very sensitive indicator for hydrogen bonding also in solution. It should be noted that hydrogen bonding can also be studied using heteronuclear NMR spectroscopy, as hetero-atoms also change their spectroscopic properties upon hydrogen bonding. For example, in case of peptides or proteins, it is an important question to define hydrogen bond donors and hydrogen bond acceptors. Solid-state NMR experiments on peptides have delineated the effects which govern the chemical shifts of carbonyl carbons, amide nitrogens, and amide protons as a function of the hydrogen bonding capabilities of different solvents [21–23]. It was found that carbonyl protonation (*e.g.* carbonyl oxygen is the hydrogen bond acceptor) causes a deshielding of the amide nitrogen [22] similar to that of the amide proton [23]. The situation was found to be more complex for the carbonyl carbon NMR chemical shift [21]. Although a low field shift of the peptide carbonyl resonance results from carbonyl protonation (H bond acceptor role), H bonding of the covalently linked NH group (H bond donor role) leads to an increased shielding of the carbonyl carbon atom. Hence, the opposite chemical shift trends caused by either NH or C=O hydrogen bonding can sometimes cancel each other and thus render the hydrogen bonding of the peptide group undetectable by chemical shift arguments. From a comparison of calculated and experimental ^{15}N chemical shift tensors of benzamide, it was concluded that chemical shift anisotropy is a very sensitive probe for hydrogen bonding [24], as the variation of the individual tensor components could be masked by measurements of the traditional isotropic chemical shift value σ_{iso}, which is defined as the trace of the chemical shift tensor.

Another very sensitive spectral probe for studying hydrogen bond properties in solution is the heteronuclear scalar coupling constant (*e.g.* $^1J_{\mathrm{NH}}$). Because a one-bond spin coupling is a through-bond phenomenon, it is a direct measure of the covalent bond character (*e.g.* hybridization, bond length). For example, *Limbach* and co-workers have used $^1J_{\mathrm{NH}}$ couplings to monitor a gradual intermolecular proton transfer between ^{15}N-pyridine and *o*-tolylic acid or 2-thiophenecarboxylic acid [25]. The $^1J_{\mathrm{NH}}$ coupling constants increased from almost 0 to about 90 Hz upon proton displacement. *Bachovchin* and co-workers [26] have used $^1J_{\mathrm{NH}}$ coupling constants (amongst other experimental parameters) to provide experi-mental evidence against the existence and catalytical importance of a putative low-barrier hydrogen bond (LBHB) in serine proteases. They measured the imidazole N–H spin coupling constants for protonated His^{57} and neutral His^{57} in the catalytic triad Asp^{102}-His^{57}-Ser^{195}. From the experimental values of 80±4 Hz for protonated His^{57} and 90±1 Hz for neutral His^{57}, they concluded that the the proton is essentially localized on $N^{\delta 1}$ in neutral His^{57} and at least 85% on $N^{\delta 1}$ when His^{57} becomes protonated and engaged in the putative LBHB. The estimation was based on the one-bond coupling constant $^1J_{\mathrm{HF}} \approx 120$ Hz for the bifluoride ion (FHF$^-$) in dipolar aprotic solvents. The coupling is about one-forth of that observed for HF (476 Hz) [27]. The decrease in the $^1J_{\mathrm{HF}}$ coupling constant in (FHF$^-$) very nicely demonstrates the sensitivity of this NMR parameter. Recently, protein-solvent hydrogen bonding was studied by $^1J_{NC'}$ coupling constants [28]. It was demonstrated that $^1J_{NC'}$ significantly increases ($^1J_{NC'} > 17\,\text{Hz}$) upon hydrogen bond formation between the carbonyl oxygen and water molecules at that site, whereas extremely low values ($^1J_{NC'} < 14\,\text{Hz}$) correspond to the absence of hydrogen bonds at the respective carbonyl sites.

Rapid tumbling motion of a molecule in solution causes an averaging of the tensorial interaction between the chemical shift tensor of a particular spin and the external magnetic field [29]. The anisotropy of the chemical shift tensor, observed as an orientational dependence of the chemical shift tensor in the solid (powder pattern) [30] is not manifested in high resolution NMR spectra in solution, but is reduced to the isotropic chemical shift, σ_{iso}. Chemical shift anisotropy (CSA), however, is responsible for an important relaxation mechanism of spin systems in solution and has been the subject of detailed investigations since the early ages of NMR [31–33]. Although experimental studies have dealt mostly with ^{19}F, ^{13}C, and ^{31}P nuclei, it is now well established that even protons can have significant and detectable CSA contributions to their relaxation [34–37]. Relaxation interference effects (*e.g.* cross-correlations), as was first pinpointed by *McConnell* [38], between CSA and dipolar couplings contain information on motional properties and chemical shift tensors. These cross-correlated relaxation mechanisms lead to a conversion of *e.g.* longitudinal *Zeeman* order $\langle I_{zk} \rangle$ into longitudinal two-spin order and $\langle 2I_{zk}I_{zl} \rangle$ [33] and can effectively be monitored by double-quantum filtered correlation spectroscopy (DQF-COSY) [36]. It also gives rise to unusual line shapes (*e.g.* differential line broadening) [33]. The effect of differential line broadening is not limited to single-quantum coherence (SQC), but a general phenomena of nuclear coherences [39], and defines the basis for a new group of pulse sequences to determine protein backbone dihedral angles [40, 41]. Specifically, in a weakly coupled AX spin system, the decay rate of the low-field component of the doublets will be faster than the decay of the upfield components, the difference being given by

$$(1/T_2)^\beta - (1/T_2)^\alpha = 16/3\, J_{AXA}(0) + 4 J_{AXA}(\omega_A) \tag{1}$$

For isotropic molecular motion, the cross-correlation spectral density function $J_{AXA}(\omega_A)$ is given by the following expression:

$$J_{AXA}(\omega_A) = 1/10(\mu_0/4\pi)\gamma_A^2\gamma_X h\langle r_{AX}^{-3}\rangle B_0 \Delta\sigma_A(\tau_c/(1+\omega_A^2\tau_c^2))1/2(3\cos^2\varphi_{AXA}-1) \tag{2}$$

where B_0 is the strength of the magnetic field, $\Delta\sigma_A$ the chemical shift anisotropy of proton A, and φ_{AXA} the angle between the unique axis of the CSA tensor σ_A and the internuclear vector r_{AX}. The CSA tensor is assumed to be axially symmetric in these equations; more general expressions have been given [33]. The cross-correlation rate $J_{AXA}(\omega_A)$ can simply be determined by monitoring the individual decay rates of the two doublet components. For ^{15}N-enriched biomolecules, more elaborate experimental schemes have been proposed [42–44]. They are based on either conversion of in-phase coherence (transverse one-spin order) into anti-phase magnetization (transverse two-spin order), or on measurements of the relative amplitudes of the two doublet components at the end of a constant-time evolution period. As was reported for O-H$\cdots$O hydrogen bonds [16], the HN proton CSA also depends strongly on the length of the hydrogen bond. For example, the measured values found in the protein ubiquitin ($MW = 8.6\,$kD) ranged from near zero in the absence of hydrogen bonding to about 14 ppm for residues involved in short hydrogen bonds. Most important, an increase in hydrogen bond strength increased the shielding parallel to the N–H bond but decreased the shielding orthogonal to this bond. As a consequence, hydrogen bonding is more sensitively probed by CSA than by measurements of the istropic chemical shift σ_{iso}. A dependence of CSA on secondary structure was observed. α-Helical amide protons have smaller CSA values (7.2$\pm$1.5 ppm) in contrast to solvent exposed amide protons (10.5$\pm$1.6 ppm) and amide protons located in β-sheets (11.2$\pm$1.6 ppm) [42].

Experimental aspects

In recent NMR-spectroscopic and structural studies [45–47] of the B_{12}-cofactor methylcobalamin (**1**) in aqueous solution, several conformational differences between the solution structure and the crystal structure [48] of **1** were observed, notably in the nucleotide loop. Conformational information obtained from ROESY data together with dihedral angle constraints obtained from three-bond homo- and heteronuclear scalar coupling constants and the spectral observation of the solvent exchangeable $R2'$-OH hydroxyl proton allowed the identification of an internally bound water molecule, linking the polar phosphate group, the amide group (N174), and the $R2'$-OH of the ribose moiety of the nucleotide loop [47]. The bound water molecule thus acts as the responsible restructuring element for the nucleotide loop of methylcobalamin in aqueous solution (Fig. 1). This detection of an exchange-labile OH-hydrogen in aqueous solution is remarkable, as precedence for such a situation is scarce [49–51]. It indicates the labile $R2'$-OH hydrogen of the ribose unit of **1** to be involved in a (*pseudo*)intramolecular hydrogen bond and to be protected in this way from exchange with the solvent. Figure 2 shows a ^{1}H NMR spectrum of methylcobalamin [45] dissolved in 90%H_2O/10%D_2O. Due to solvent presaturation, the $R2'$-OH hydroxyl ribose proton signal was barely observed at 5.5 ppm. Selective excitation resolved the signal as a doublet with 4.5 Hz splitting (Fig. 3). Up to now, only a few $^3J_{\mathrm{(HOCH)}}$ coupling constants have been reported, and a *Karplus* relationship for a similar system is not available in the literature. The observed coupling constant of 4.5 Hz for $^3J_{\mathrm{(HOCH)}}$ allows for an additional dihedral angle constraint with respect to the spatial orientation of the $R2'$-OH hydroxyl group consistent with a structure in which O–H points roughly towards H(N174) and suggests a *gauche* conformation (*ca.* 60°) rather than an angle close to the *Karplus* maxima (180° or 0°) or near the *Karplus* minimum (90°). The decay rates of the two doublet components of the hydroxyl hydrogen $R2'$-OH in methylcobalamin were determined by a standard *Carr-Purcell-Meiboom-Gill* (CPMG) T_2-pulse sequence [52], but using multiplet selective excitation [53] and inversion pulses [54] (see legend of Fig. 3). Figure 3 shows the decay of the two doublet components. From the differential decay of the two doublet components, the cross correlation rate was determined to be about 1.1 Hz using Eq. (1). The

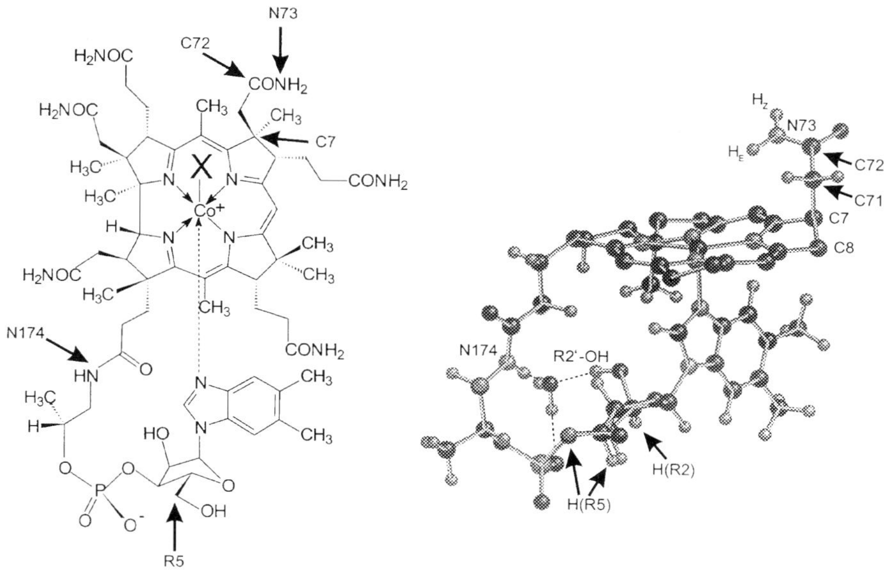

Fig. 1. (Left) Structural formula of methylcobalamin (**1**, $X = CH_3$) and the aquocobalamin cation (**2**, $X = H_2O^+$); (Right) View of relevant parts of the solution structure of **1**, showing the corrin macrocycle, the nucleotide loop, and the c-acetamide side chain, highlighting the internal, tetrahedrally coordinated water, the hydroxyl hydrogen $R2'$-OH, and the amide proton H(N174). The dashed lines represent hydrogen bonds

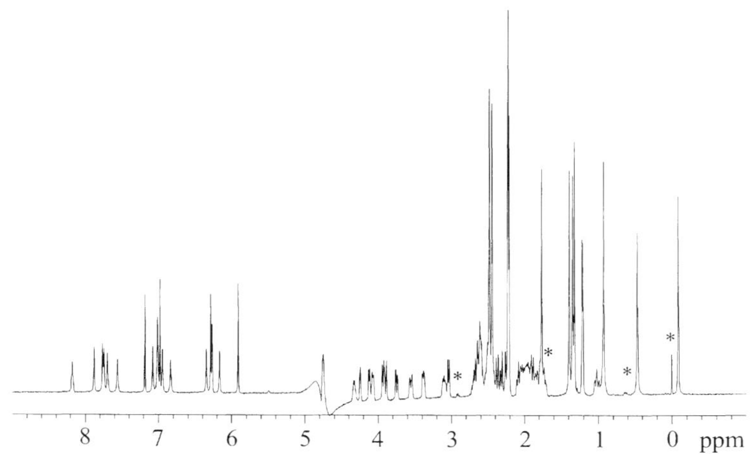

Fig. 2. 500 MHz ^{1}H NMR spectrum of **1** dissolved in 90% H_2O/10% D_2O

chemical shift anisotropy $\Delta\sigma_A$ was calculated to be about 28 ppm from Eq. (2) (assuming axial symmetry of the magnetic shielding tensor) and taking the dihedral angle φ_{AXA} and the distance r_{AX} from the solution structure of methylcobalamin where A is $R2'$-OH and X is H($R2$), respectively. The overall reorientational correlation time τ_c was determined to 250 ps based on ^{13}C-T_1 values determined by standard inversion-recovery experiments. This proton CSA value is very close to

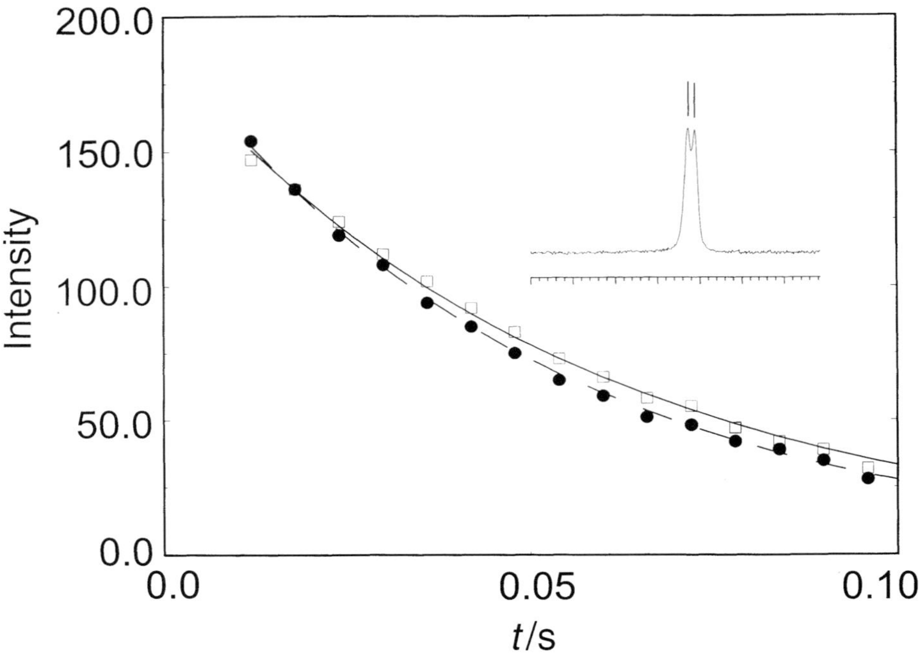

Fig. 3. T_2-decay curves for the doublet components of $R2'$-OH in methylcobalamin. Selective T_2-measurements were performed using a 30 ms 270° *Gaussian* excitation pulse [53] and 50 ms G3 *Gaussian* cascade inversion [54] RF pulses. From the differential decay of the two doublet components, a chemical shift anisotropy - dipolar cross correlation rate (CSA-DD) of 1.1 Hz was determined. Using Eq. (2), the chemical shift anisotropy of the $R2'$-OH proton was calculated to
$$\Delta\sigma = 28\,\text{ppm (see text)}$$

values obtained from proton anisotropic chemical shift spectra obtained in a single crystal of hexagonal ice (28.5 ppm) [55] and proton NMR powder spectra of ice, 34.2 ± 1.0 [56] and 34 ± 4 ppm [57]. Theoretical calculations studying the influence of secondary hydrogen-bonding effects on the shielding of hydrogens in ice also confirmed both a CSA value of about 30 ppm and the axial symmetry of the magnetic shielding tensor of a hydroxyl proton involved in a hydrogen bond [58]. Thus, the hydrogen bond formed between the $R2'$-OH and the internal, tetrahedrally coordinated water molecule is well defined in solution and comparable to hydrogen bonds formed between water molecules in the solid state.

H/D Isotope Fractionation Factor

The isotopic fractionation factor φ of an exchange-labile proton in a molecule is defined as the equilibrium constant for the exchange process of this particular proton with deuterons from the solvent:

$$XH + DS(\text{bulk solvent}) \rightleftharpoons XD + HS(\text{bulk solvent})$$

$$\varphi = \frac{[XD][HS]}{[XH][DS]}$$

A value of 1 reflects an equal distribution of protons and deuterons between the exchange-labile position X and the solvent S; a φ value >1 indicates a preference for D over H on the position X, whereas a value <1 indicates a preference for H over D. The equilibrium constant can be analyzed in terms of relative zero-point energies of the X-H and X-D bonds [3, 59–61]. Fractionation factors can be rationalized qualitatively by considering effective one-dimensional potentials or effective force constants. In case that the exchange-labile site X has a stronger force constant for the bond X-[H, D] than for the solvent S, the fractionation factor will be >1 and *vice versa*. In the case of a hydrogen bond X-H$\cdots Y$, the force constant of the X-H bond is decreased resulting in a fractionation factor < 1. There is extensive literature about deuterium fractionation factors for low molecular weight compounds such as alcohols, phenols, carboxylates, hydroxyls, imidazoles, amines, and amides, most of them being close to 1 in aqueous solution [62–64]. Deviations occur for hydrogen bonded systems. For example, in weakly bound solute-solvent complexes of water-dioxane, water-methanol, and tetrahydrofuran-fluoroform, fractionation factors > 1 have been observed. In contrast, compounds that form strong hydrogen bonds typically exhibit low fractionation factors ($\varphi < 1$), *e.g.* F_2H^- in water ($\varphi = 0.6$), dimers of 4-nitrophenolate ($\varphi = 0.31$), trifluoroacetate ($\varphi = 0.42$), 3,5-dinitrobenzoate ($\varphi = 0.30$), 3,5-dinitrophenolate ($\varphi = 0.36$), and pentachlorobenzoate ($\varphi = 0.40$) (all in acetonitrile) [65]. The larger values of φ in protic solvents is presumably due to competing hydrogen bonding with the solvent. Until recently, spectral complexity confined the measurements of isotopic fractionation factors to low molecular weight compounds. Exceptions were measurements on *e.g.* the hydrogen shared between the two glutamate residues E168 and E211 in enolase and that between glutamate E217 and N1 of bound adenosine in adenosine deaminase, each having φ values of about 0.4 [66–68]. In this case, the particular low φ values have been attributed to low-barrier hydrogen bonds, and the importance of this type of hydrogen bonding to enzyme catalysis has been discussed by *Cleland* [66]. Recently, a new NMR method was introduced [69, 70] which allows for the determination of fractionation factors in uniformly ^{15}N-labeled proteins. The method relies on measurement of the intensities of ^{15}N-^{1}H correlations in a range of H_2O/D_2O solvent mixtures. This technique has been applied to staphylococcal nuclease both in its unligated state and in its ternary complex with Ca^{2+} and the inhibitor thymidine $3',5'$-bisphosphate [70]. The fractionation factors varied between 0.34 (threonine T120) and 1.42 (glycine G55), with average values and standard deviations of 0.84 ± 0.19 and 0.86 ± 0.17 for unligated and ligated nuclease. α-Helical regions displayed values of about 0.79 ± 0.10, whereas slightly larger φ-values (0.91 ± 0.15) were found for residues located in β-sheet secondary structures. No clear correlation was found between fractionation factors and N–O distances for the amide-amide hydrogen bonds observed in the crystal structure of unligated nuclease. This reflects the subtle interplay of mechanisms responsible for determining fractionation factor in complex molecules (*e.g.* off-line bending motions in addition to in-line stretching vibrational oscillations around the equilibrium position). Additionally, conformational differences between the solid and the liquid state of staphylococcal nuclease could not be ruled out. However, the significantly reduced value ($\varphi = 0.34$) for the backbone amide of threonine T120 was attributed to a hydrogen bond formed with the charged side-chain carboxylate group of aspartate D77. Thus, in accordance with data obtained in the gas-phase, hydrogen bonds involving a charged acceptor site were found to be energetically more favourable and displayed decreased fractionation factors. In general, the study showed that, on average, protium is accumulated to the greatest extent in α-helices and at sites of cooperative hydrogen bonds involving charged moieties, in contrast to the enrichment of deuterium seen in more unstructured parts of the proteins (*e.g.* loop regions).

Experimental aspects

Although fractionation factors φ can be obtained from the resonance intensities (*i.e.*, the protium level) in a single spectrum of known intermediate solvent composition (for example 50% D_2O), it is more accurate to determine the proton occupancies at a particular molecular site as a function of solvent composition. Fractionation factors can be obtained by linear least-squares analysis of

$$1/y = C(\varphi(1-x)/x + 1) \tag{3}$$

where y is the peak area (intensity), x is the mole fraction of H_2O, and C is a normalization factor [70]. Of course, it is crucial that equilibrium has been reached before spectra are acquired. If there is no spectral overlap, one-dimensional ^1H NMR spectra are sufficient to obtain the necessary information. To obtain absolute intensities as a function of solvent composition, the signals can either be referenced to an external or internal standard (*e.g.* other non-exchangeable protons of the molecule). We have recorded ^1H spectra of methylcobalamin (**1**) (see Figs. 1 and 2) at different solvent composition, and the signal intensities of $R2'$-OH as a function of H_2O mole fraction is shown in Fig. 4. The fractionation factor for

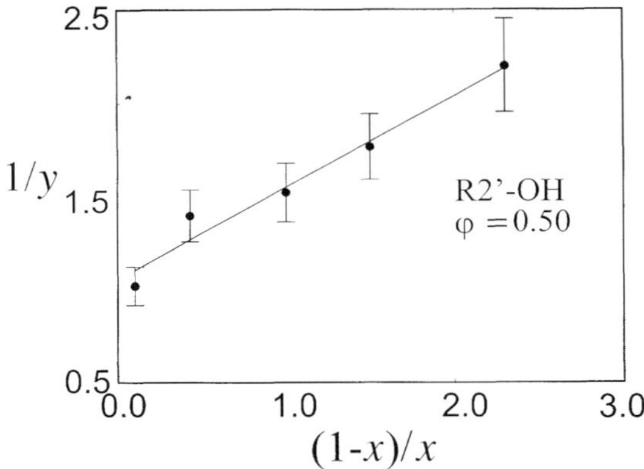

Fig. 4. Measurement of H/D fractionation of $R2'$-OH in methylcobalamin. Plots of normalized ^1H NMR signal intensity (yC) as a function of the mole fraction H_2O (x) according to the equation [69,70] $(yC)^{-1} = (\varphi(1-x)/x)+1$, where y is the signal intensity, C is a normalization factor, and the slope of the line is the H/D fractionation factor (φ). The signal intensity y was obtained by referencing to the non-exchangeable methyl proton signal which resonates most upfield in the ^1H NMR spectrum (0.5 ppm). To minimize signal attenuation of the exchanging $R2'$-OH signal due to exchange with bulk water, selective excitation of the $R2'$-OH signal (50 ms, self-compensating *Gaussian* excitation pulse [53]) was applied

$R2'$-OH was determined to be 0.5 (according to Eq. (3)), typical for an exchangeable proton involved in a strong hydrogen bond, and corroborates the existence of the hydrogen bond formed to the internally coordinated water molecule as delineated from NOEs and coupling constant information (see below).

The dimensionality of the experiment and thus the spectral resolution can be increased by recording a ^{1}H-^{15}N correlation spectrum [69, 70], or, if a doubly labeled protein is available, a HA(CA)CO spectrum [71]. In this experiment, the ^{1}H$^\alpha$ resonance of residue i is correlated with the intraresidue carbonyl ^{13}C$'$ resonance. If the sample is equilibrated in 50% H_2O/50% D_2O, the ^{1}H$^\alpha$-^{13}C$'$ correlation will show a splitting in the ^{13}C$'$ dimension as a result of the two-bond isotope shift, corresponding to protonated and deuterated states of the directly attached nitrogen. Thus, there is no need for external referencing, and very accurate data can be obtained on a sample dissolved in a solvent mixture of 50% H_2O/50% D_2O. However, if no isotope labeling is available or if rapidly exchanging and hence unresolved hydroxyl protons (rapidly exchanging protons experience a chemical shift similar to bulk water) are under scrutiny, we have devised an alternative method to study fractionation factors in aqueous solution. The experimental scheme is basically a 2D ROESY [72] sequence with standard modifications to ensure water suppression using pulsed field gradients according to the WATERGATE technique [73]. In the 2D ROESY experiment, cross peaks are found at the water chemical shift in the indirect dimension and non-exchangeable protons H_{non} in the direct dimension. This transfer is caused by a relay mechanism [74] which depends on the nature of magnetization transfer operative during the spin lock period, consisting of both TOCSY-type transfer due to scalar coupling (^nJ(OH-H_{non})) [74] and dipolar interaction (σ_{ROE}). The flow of magnetization for this process is as follows:

$$\boldsymbol{I}_x(H_2O) \rightarrow (k_{ex}) \rightarrow \boldsymbol{I}_x(OH) \rightarrow {^nJ}(OH - H_{non})/\sigma_{ROE} \rightarrow H_{non} \qquad (4)$$

The intensities of these exchange-relayed cross peaks are governed by the transfer function (scalar and/or dipolar coupling) and the spin densities of the protons involved. The transfer function and the spin densities of the non-exchangeable protons do not change with solvent composition. In contrast, the spin densities of exchangeable protons (*e.g.* hydroxyl protons) are a function of the H_2O/D_2O ratio. Hence, measuring cross peak intensities for a range of H_2O/D_2O solvent mixtures gives an estimate of the spin densities of the exchangeable protons or, in other words, the H/D isotope fractionation factor φ. We have again used the corrinoid cofactor methylcobalamin (**1**) as an example. A recent NMR study has shown that the solution structure is considerably different from the crystal state, mainly caused by the existence of an internal tetrahedrally coordinated water molecule. The hydrogen bonding network of this water molecule comprised the secondary ribose hydroxyl hydrogen ($R2'$-OH), which significantly retarded the intermolecular exchange with bulk water and allowed the NMR observation of this hydroxyl hydrogen at 5.50 ppm. The hydrogen of the primary ribose hydroxyl group ($R5'$-OH) was still unobservable due to rapid exchange. Figure 5 shows a trace taken from a 2D ROESY spectrum of methylcobalamine along the water resonance in ω_1. Apart from negative cross peaks which are due to intermolecular ROEs between hydration water molecules and protons of methylcobalamin, there are four

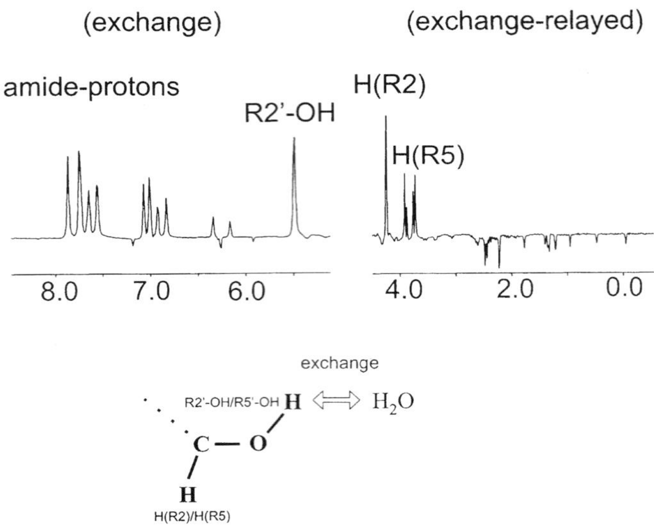

Fig. 5. Cross section parallel to the ω_2-axis taken at the ω_1 frequency of the water resonance through a 200 ms WATERGATE-ROESY of methylcobalamin (**1**). Water NOEs found for protons located near the bulk water hydrations sites occur as negative cross peaks in the spectrum. Positive cross peaks belong to exchange cross peaks with the bulk water signal ($R2'$-OH at 5.50 ppm) or exchange-relayed cross peaks (H($R2$), H($R5$a), H($R5$b)), typical of protons located in the vicinity of exchangeable protons [74]

interesting positive cross peaks: $R2'$-OH, H($R2$), and the two diastereotopic protons H($R5$a) and H($R5$b). The cross peak between bulk water and $R2'$-OH is caused by intermolecular exchange, whereas the others are exchange-relayed peaks, presumably due to TOCSY-type transfer. For the determination of φ, the general procedure of *Loh* and *Markley* was used [69, 70]. Four samples with varying amounts of H_2O/D_2O were prepared; the mole fractions (x) of H_2O were 0.9, 0.7, 0.5, and 0.3. The spin densities of the exchangeable protons were determined by measuring the relative ROESY cross peak intensities between a pair of non-exchangeable protons and a cross peak involving the exchangeable proton of interest (see legend of Fig. 6). Figure 6 shows a linearized plot of the intensities *vs.* H_2O mole fraction according to Eq. (3). Due to the resolved resonance of $R2'$-OH, both, direct ROESY cross peaks between $R2'$-OH and non-exchangeable protons (H($R2$)) as well as exchange-relayed cross peaks (detected at H($R2$) at the bulk water trace in ω_1) could be analyzed. Quantitative interpretation of the measured spin densities as a function of solvent composition was achieved by calibration with one-dimensional ^{1}H-spectra (see above). This was necessary because the absolute values of the experimentally determined spin densities are additionally governed by magnetization exchange between the solvent and the solute, a process which itself is a function of the deuteration level. For example, solvent presaturation and/or the use of an interscan delay shorter than the long solvent T_1 attenuates resonances of protons in rapid exchange with solvent. Direct and

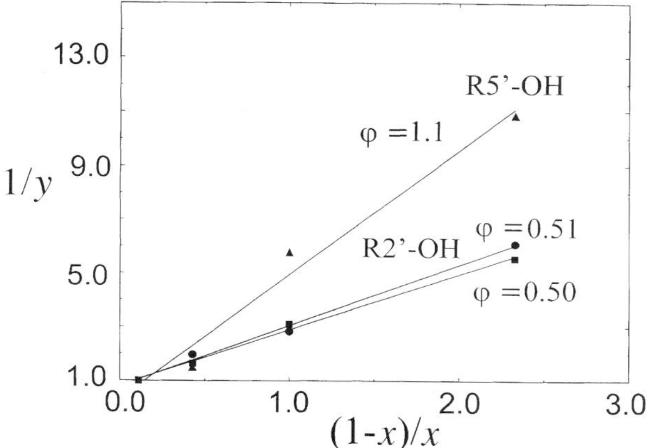

Fig. 6. Measurement of H/D fractionation of $R2'$-OH in methylcobalamin based on a two-dimensional ROESY experiment. Plots of normalized cross peak intensity (yC) as a function of the mole fraction H_2O (x) according to Refs. [69,70]. The signal intensities y for $R2'$-OH were obtained from referencing the direct (exchange, shown as squares) bulk water ROESY cross peak of $R2'$-OH to ROESY cross peaks of $R2'$-OH to non-exchangeable protons H($R2$). A second set of data for $R2'$-OH was obtained using the exchange-relayed cross peak of H($R2$), as taken from the water trace (Fig. 5) and again referencing to a cross peak involving a non-exchangeable proton (shown as circles). The H/D fractionation of $R5'$-OH was determined only from the exchange-relayed cross peak to both H($R5$) protons

exchange-relayed magnetization transfer results in identical values for φ, thus proving the reliability of the method (Fig. 6). The φ values for $R2'$-OH obtained with the three methods discussed were 0.50 (one-dimensional ^{1}H-based method, reference value; Fig. 4), 0.50 (direct transfer), and 0.51 (exchange-relayed transfer). The significantly reduced value of φ is due to hydrogen bond formation to the internally bound water molecule. For the rapidly exchanging $R5'$-OH proton a fractionation factor of about 1.1 was obtained using the relayed magnetization transfer. This is expected for exchangeable protons which are not involved in strong hydrogen bonds. In solution, the $R5'$-OH of methylcobalamine points towards the solvent and is freely accessible to bulk water molecules. Similar values were found for low molecular weight alcohols and hydroxyls in aqueous solution [62–64]. This shows that even for rapidly exchanging hydroxyl protons the 2D ROESY experiment provides unique access to φ values which otherwise cannot be obtained in aqueous solution.

Intermolecular XH Exchange

Hydrogens bound to N, O, and S atoms of polar groups within a solute molecule dissolved in water are in continual exchange with the hydrogens of the solvent. In contrast, carbon-bound hydrogens do not exchange easily. Although polar-group hydrogens (XH) are generally exchange-labile, they are covalently bound and exchange with solvent hydrogens only as a result of distinct chemical reactions. The underlying catalytic reaction steps are reversible proton transfer reactions between donor and acceptor groups. The primary step in a proton transfer reaction can be regarded as a diffusion-limited collision and hydrogen bond formation between a proton donor (AH) and a proton acceptor (B), leading to a so-called encounter complex. The next event is a proton redistribution across the hydrogen bond, apparently *via* quantum mechanical tunneling [76].

$$AH + B \underset{k_{-D}}{\overset{k_D}{\rightleftharpoons}} (AH\text{---}B \overset{K_C}{\rightleftharpoons} A\text{---}HB) \underset{k_D}{\overset{k_{-D}}{\rightleftharpoons}} A + HB$$

The equilibrium constant for this reversible proton redistribution is given by

$$K_c = 10^{\Delta pK} \tag{5}$$

where

$$\Delta pK = pK_B - pK_A \tag{6}$$

For a successful proton transfer reaction, the reaction rate k_{tr} is given by

$$k_{tr} = k_D(10^{\Delta pK}/10^{\Delta pK} + 1) \tag{7}$$

where k_D is the second order rate constant for the diffusion-limited collision. Hence, the rate-limiting step for a hydrogen exchange reaction is the proton transfer reaction from the proton donor (acid) to the proton acceptor (base). Thus, k_{ch}, the intrinsic chemical exchange rate characteristic of the catalyst (Cat), can be set equal to the limiting proton transfer rate constant k_{tr} (Eq. (7)). In general, more than one catalyst can be present, and the resulting first order exchange rate constant (k_{ex}) is given as the sum of all possible contributions.

$$k_{ex} = \Sigma\, k_{ch}[Cat] \tag{8}$$

Two cases can be distinguished: the transfer of a proton proceeding either energetically downhill (from a stronger to a weaker acid, $\Delta pK > 1$) or uphill (from a weaker to a stronger acid, $\Delta pK < 1$). In the first case, every collision leads to a successfull proton transfer, and transfer proceeds at the maximum diffusion-limited rate defined by k_D. For example, nucleic acid NH ring protons or polar protein side chain protons (*e.g.* OH) have pK-values much lower than 15; therefore, exchange of these protons is very effectively catalyzed by the hydroxyl ion. Examples for the latter case are peptide amide protons. The deprotonation pK for a peptide group CONH is about 18.5 [76]. As a consequence, proton exchange of peptide amide groups is dominated by base catalysis down to about $pH = 3$. The first order exchange rate constant (k_{ex}) (assuming OH^- and H^+ being the only catalysts) is thus given by

$$k_{ex} = k_{OH}[OH^-] + k_H[H^+] \tag{9}$$

k_{OH} and k_H are defined according to Eq. (7). In the case of protein amides, this results in a V-shaped $\log(k_{ex})$ *vs.* pH profile with a minimum rate occurring between pH 2 and 3. This only holds for small molecules. In contrast, in more complex molecules (*e.g.* proteins, nucleic acids, protein complexes), a variety of factors can influence the exchange rates. Nearest neighbour groups can impose inductive (electron withdrawing) or electrostatic effects that shift the minimum of the pH profile along the pH axis. Steric effects can modify the accessibility and, hence, the second order rate constant for the formation of the encounter complex. For peptide group amide protons, the important factors have been accurately calibrated in small molecule models [77–79]. Thus, it is now possible to reliably predict intermolecular exchange rates for random-coil polypeptides (and also polynucleotides) under ambient conditions. The most important influence imposed by the structure of the molecule, however is caused by a possible engagement of the exchangeable proton in a hydrogen bond. Proton transfer always requires, a an intermediate step, the formation of a hydrogen bond between donor and acceptor (exchange catalyst and, for example, protein amide proton). If the donor proton or the acceptor site is already involved in a pre-existing intramolecular hydrogen bond, the transfer will be impeded, and exchange significantly slowed down. This has already been reported in *Eigen's* classical work on proton transfer theory [80]: the hydroxylic proton of salicylate is removed by $^-$OH three orders of magnitude slower than the diffusion-limited rate, characteristic for non-hydrogen bonded hydroxyl protons. Many other examples exist which display slowing factors of up to almost six decades [81, 82] and demonstrate that significant hydrogen bonding can be found also in small molecules. Protein hydrogen exchange studies were initiated by *Linderstrøm-Lang* in the mid-1950s [83, 84] and have played a major role in the description of protein stability, protein folding and unfolding events, and in the characterization of the dynamical properties (*e.g.* structural fluctuations) of proteins ever since (for broad background reviews on this subject, see Refs. [60, 76, 85]). It is generally assumed that protection of amide protons in proteins from exchange with solvent protons involves hydrogen bonding. The exchange of protected protons occurs through structural fluctuations, leading to a transient severing of the blocking hydrogen bond. Exchange of the peptide amide proton with solvent protons can only occur in the open state, in which the hydrogen bond has been broken. In the closed state, the proton is still involved in the hydrogen bond and cannot form the rate-limiting hydrogen bond to the exchange catalyst. However, the open state can differ from the closed state by more than the hydrogen bond alone. As examples, neighbouring secondary structure segments may unfold in a cooperative manner or, alternatively, the open state can be differently stabilized by additional favourable interactions. Thus, the probability of the open state will depend not only on the free energy of the hydrogen bond but rather on both the hydrogen bonding energy of the closed (native) state and the interactions forming in the open state. The kinetic and thermodynamic relationships that relate structural isomerization (the breakage of a blocking hydrogen bond) with measured exchange rates (k_{ex}) were first formulated by *Berger* and *Linderstrøm-Lang* [86]. More general formulations have been given by *Hvidt* and *Nielsen* [60].

The observed exchange rate k_{ex} is given [76] by

$$k_{\mathrm{ex}} = k_{\mathrm{op}}k_{\mathrm{ch}}[Cat]/(k_{\mathrm{cl}} + k_{\mathrm{ch}}[Cat]) \qquad (10)$$

in which k_{op} and k_{cl} are the rate constants for the breakage and forming of the blocking hydrogen bond. k_{ch} is the intrinsic chemical exchange rate of the catalyst (*Cat*). Two cases can be distinguished. If refolding of the structural opening is fast compared to the intrinsic exchange rate ($k_{\mathrm{cl}} \gg k_{\mathrm{ch}}[Cat]$), exchange will be a second order reaction dependent on the concentration of the catalyst ($k_{\mathrm{ex}} = K_{\mathrm{op}}k_{\mathrm{ch}}[Cat]$; $K_{\mathrm{op}} = k_{\mathrm{op}}/k_{\mathrm{cl}}$). This limiting case is also known as the EX2 limit [76,87]. From the value of K_{op} (determined by $K_{op} = k_{ex}/k_{ch}[Cat]$), the free energy stabilizing the closed (native) state can then be determined from the expression $\Delta G = -RT \ln K_{op}$. In contrast, the EX1 limit is reached when reforming of the hydrogen bond is slower than the intrinsic intermolecular exchange with solvent protons. The exchange rate is then equal to the opening rate k_{op} of the structural isomerization process (the breakage of the hydrogen bond).

Although a large number of different techniques for hydrogen exchange measurements exist [88–92], the ultimate resolution of measuring hydrogen exchange at individual atomic sites is only offered by neutron diffraction [93] and by NMR methods. In this review, we limit our discussion to NMR based methods. In principle, three fundamentally different NMR methods exist. Firstly, hydrogen exchange can be probed by so-called exchange-out techniques [88]. The intensity of the resonance line of a particular proton is proportional to the proton (^{1}H) concentration at this respective site. The time dependence of the line intensity after exposure to D_2O is a direct measure of the exchange rate at that particular site. If the signals are not separated, two-dimensional homonuclear or heteronuclear NMR experiments have to be applied. This type of experiments have successfully been applied to proteins and protein complexes [94–100]. A most striking application of this technique was the identification of an antibody binding site on Cytochrome c, defined by hydrogen exchange rates combined with the resolution obtained from a two-dimensional ^{15}N-^{1}H correlation spectrum [101]. Secondly, hydrogen exchange rates can be measured by saturation transfer techniques [102], where attenuation of the exchangeable proton signal is monitored after extended irradiation of the water signal. In its simplest form, the exchange rates of rapidly exchanging protons are measured from the effect of presaturation on simple one-dimensional spectra [103–105]. This requires the recording of two one-dimensional spectra in H_2O with and without presaturation of the water resonance. An extended version of the original experiment and applicable to ^{15}N-labeled proteins was introduced [106]. A third class of experiments is based on dynamic NMR spectroscopy [107]. Chemical exchange with bulk water protons contributes an additional relaxation mechanism to the transverse relaxation of the exchangeable proton. By measuring the transverse relaxation rate and accounting for non-exchange contributions, the exchange rate can be measured.

Experimental aspects

We have recently determined the crystal structure and the solution structure of aquocobalamin by NMR spectroscopy [108] (see Fig. 1). The NMR data confirmed the crystallographically determined occupation of the axial coordination site at the Co(III) center by water, as well as the occurrence of an intramolecular hydrogen bond to the axially coordinating water molecule in solution as observed in the crystal structure of the aquocobalamin ion. However, significant differences of the structure of aquocobalamin in the crystalline state and in aqueous solution were indicated from NOE data concerning the time-averaged conformation of the hydrogen bonding c-acetamide side chain ($H_{E/Z}$N73). The H-bonding c-acetamide side chain exhibits a *syn-clinal* arrangement of the bonds C8–C7 and C71–C72 (Fig. 1), whereas in the crystal structure these bonds are nearly *anti-periplanar*. In the crystal structure of aquocobalamin, the (less basic) lone pair, oriented *trans* to the amide nitrogen, accepts the hydrogen bond. Hence, in solution the carbonyl oxygen lone pair oriented *cis* to the amide nitrogen is indicated to act as the hydrogen bond acceptor for the cobalt coordinated water proton in solution. We have investigated the chemical exchange behaviour of the amide proton $H_E(N73)$ and could unambigiously demonstrate the engagement of this amide group in an intramolecular hydrogen bond to the cobalt coordinated water. Intermolecular exchange rates with bulk water protons were measured using a saturation transfer based NMR method [102–105]. To this end, we have recorded two sets of one-dimensional ^1H spectra (with and without presaturation of the water resonance) as a function of *pH*. In the absence of cross relaxation, attenuation of the amide resonance in the experiment with presaturation depends on the amide hydrogen exchange rate, k_{ex}, in the following manner [105]:

$$k_{ex} = (1 - M_{ps}/M_0)/T_{1app} \qquad (11)$$

Here M_{ps} is the resonance intensity in the experiment with presaturation, M_0 is its intensity without presaturation, and T_{1app} is the apparent longitudinal relaxation rate as measured in a selective T_1 experiment. T_{1app} includes the effects of both exchange and true relaxation. Eq. (11) can be rewritten as

$$k_{ex} = (M_0/M_{ps} - 1)/T_1 \qquad (12)$$

where T_1 describes the true longitudinal relaxation as measured in a selective T_1 experiment in the absence of chemical exchange. Typically, this rate is dominated by the cross-relaxation to other protons. For small compounds, homonuclear cross-relaxation is typically much smaller than the intrinsic longitudinal relaxation rate and therefore may be safely neglected. For proteins, however, the k_{ex} values as determined from Eq. (12) may include a substantial contribution from magnetization exchange caused by cross-relaxation. Examination of the *pH vs.* exchange rate profile (Fig. 7) for hydrogen exchange of amide protons with bulk water indicated the c-acetamide group $H_E(N73)$ to exchange its protons slower at a *pH* smaller than *ca.* 3.5 but faster at a *pH* above about 3.5 than, for example, the acetamide function attached to the a-acetamide side chain $H_E(N23)$. Thus, an increased resistance against acid catalyzed hydrogen exchange, as well as an increased rate of apparent base catalyzed hydrogen exchange, can be determined for the c-acetamide function. This apparent acidification of the c-acetamide function can be rationalized by the existence of a specific hydrogen bond involving its carbonyl oxygen. The existence of the H-bond between the cobalt-coordinated water molecule and the c-acetamide carbonyl oxygen is also supported by arguments based upon ^{15}N-chemical shifts (see above).

The application of dynamic NMR spectroscopy to studies of hydrogen bonding is exemplified by NMR investigations of the $R2'$-OH hydroxyl proton in methylcobalamin [45–47]. The measurement (see legend of Fig. 8) of T_2 relaxation and (from there) of the exchange rate k_{ex} of the $2'$-OH hydrogen *via* NMR in the ribose moiety of methylcobalamin as a function of *pH* provided the intermolecular exchange profile shown in Fig. 8. The experimental data could be simulated with a non-linear least-squares fit according to Eq. (9). This result indicates the prevalence of the so-called EX2-limit [76, 87], implying that the

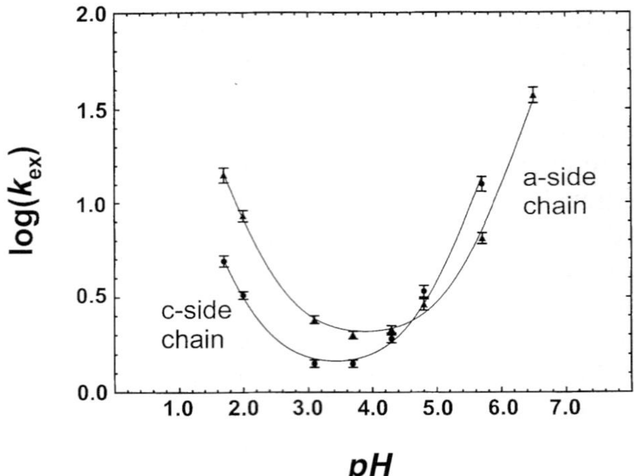

Fig. 7. Amide hydrogen bonding probed by intermolecular exchange measurements. *pH* profile of the intermolecular proton exchange rates (s^{-1}) of the trans (H_E) amide protons of the two acetamide (a, H_E(N23); c H_E(N73)) functions in aquocobalamin [107]. The experimental data for each acetamide were fitted by a hyperbola according to Eq. (9). Exchange rates were determined using saturation transfer [101] as described in the text

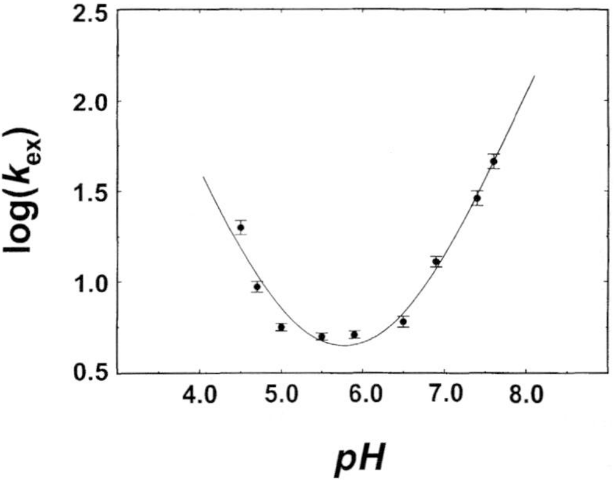

Fig. 8. Intermolecular exchange profile of *R2′*-OH in methylcobalamin [45–47]. The exchange rates k_{ex} are given in s^{-1}; solid curves were drawn using Eq. (9). NMR solutions: 10 mM solutions, sample size 0.7 cm³, buffered with 8 mM, 10 mM, 15 mM, 40 mM, 70 mM and 100 mM phosphate buffer solutions, *pH* 5.0, as well as 10 mM solutions, sample size 0.7 cm³, buffered with 100 mM phosphate buffer in the *pH* range from 4.5 to 7.6; 26°C. Intermolecular exchange rates (k_{ex}) were determined from selective T_2-measurements based on the equation $1/T_2 = 1/T_2$(dipolar)$+k_{ex}$. Selective RF pulses: excitation: 30 ms 270° *Gaussian* excitation pulse [53]; inversion: 50 ms G3 *Gaussian* cascade [54]. The T_2-values were corrected for the dipolar contribution to the linewidth $1/T_2$ (dipolar) and exchange contributions due to phosphate catalyzed intermolecular exchange. The dipolar contribution to the linewidth was determined to be 2.0 Hz by extrapolating the T_2-value to zero concentration of both catalysts, (hydrogen phosphate (HPO_4^{2-}) and hydroxyl ion (OH^-)

establishment of the internal equilibrium is much faster than the external exchange processes. From analysis of the exchange rate *vs. pH* profile according to Eq. (10) and using Eq. (7) for calculation of the intrinsic exchange rate, the equilibrium constant K_{eq} for reversible hydrogen bond formation of the $R2'$-OH hydrogen was calculated to $K_{eq} = 88$. From the value of K_{eq} determined this way, the *Gibbs* free energy of protection (ΔG_{prot}) was then determined from the expression $\Delta G_{prot} = -RT\ln K_{eq}$, yielding a value for ΔG_{prot} of $-11.3\,\mathrm{kJ\cdot mol^{-1}}$. The *Gibbs* free energy of protection corresponds to the difference between the *Gibbs* free energy of a state ("protected state") in which the $R2'$-OH proton acts as a hydrogen bond donor to the bound water molecule, and an "unprotected" state in which the protecting hydrogen bond is broken and in which the $R2'$-OH proton forms a weaker hydrogen bond to a bulk water acceptor molecule. In this "unprotected" state, the $R2'$-OH proton is accessible to the exchange catalyst (*e.g.* OH$^-$) and the bulk water. Our result for the free energy of protection ($\Delta G_{prot} = -11.3\,\mathrm{kJ\cdot mol^{-1}}$) of a ribose $R2'$-OH significantly differs from a value previously obtained in wet *DMSO* solutions [109]. For cyclic nucleotide monophosphates, free energies of protection by an assumed water molecule were obtained there whose magnitudes were considerably smaller ($-6.3\,\mathrm{kJ\cdot mol^{-1}}$) than the values obtained by us. This difference emphasizes the importance of the solvent as H-bond partner with respect to the strength of the H-bond. Our value of $\Delta G_{prot} = -11.3\,\mathrm{kJ\cdot mol^{-1}}$ compares favorably with the recently determined effect of a point mutation of a single nucleotide to a 2'-deoxynucleotide upon the binding of *RNA* substrate in the catalytic core of the *Tetrahymena* ribozyme [110]: the major contributor to the increased stability of the *RNA* complex (compared to that of the mutant) could be traced back to the $R2'$-OH of a critical nucleotide residue and N1 of an adenosine. The elimination of a specific $R2'$-OH hydrogen bond donor by substituting a ribonucleotide by a 2-deoxyribonucleotide leads to a 33-fold increase in the dissociation constant (K_d) of oligonucleotide binding to *Tetrahymena* ribozyme, representing a difference in free energy ($\Delta\Delta G$) of $9.2\,\mathrm{kJ\cdot mol^{-1}}$. This value was suggested to correlate closely with the free energy of the H-bond between a $R2'$-OH and the adenosine N1. Information about the energetics of hydrogen bonding interactions in biomolecules in aqueous solution, although highly desirable [111], is scarce [3, 4]. Deletions (*e.g.* by point mutations) of hydrogen bonding groups in proteins [112] and nucleic acids [113], often where found to lead to very small changes in the thermodynamic stability, presumably by replacing a direct hydrogen bond by a water-mediated hydrogen bond, or by changing one or both hydrogen bonding partners. Such compensating structural effects make it thus very difficult to determine accurately the strength of individual hydrogen bonds simply by evaluating the free energy difference caused by the deletion of a particular hydogen bond. Thus, dynamic NMR studies provide unique experimental access to hydrogen bonding energies in aqueous solution.

Conclusion

Increases in our knowledge of and our ability to experimentally measure hydrogen bond induced changes in H/D isotope fractionation, proton chemical shift anisotropy (CSA), and intermolecular exchange processes with bulk water molecules provide a rich avenue to analyze systems in aqueous solution and acquire information by NMR spectroscopy not generally accessible by other means. Recent work has shown how detailed information on hydrogen bonded protons can be obtained by these solution NMR approaches. It is anticipated that additional theoretical work will provide physical insight into the correlations between H/D isotope fractionation, proton chemical shift anisotropy, and hydrogen bond energies and that these NMR approaches will advance our current understanding of hydrogen bonding and its impact on the structural propensities of molecules containing exchange labile hydrogens.

Acknowledgements

This work was supported by the *Austrian National Science Foundation* FWF (Project No. P10816-GEN and P11600-GEN). We are grateful to *Hoffmann-La Roche*, Basel, Switzerland, for a generous gift of vitamin B_{12}.

References

[1] Early consideration of a "hydrogen bond": a) Moore TS, Winwill TF (1912) J Chem Soc **101**: 1635; b) Latimer WM, Rodenbush WH (1920) J Am Chem Soc **42**: 1419

[2] Pauling L, Corey RB, Branson HR (1951) Proc Natl Acad Sci **37**: 205; Pauling L (1960) The Nature of the Chemical Bond, 3rd edn. Cornell University Press, New York

[3] Hibbert F, Emsley J (1990) Adv Phys Org Chem **26**: 255

[4] Jeffrey GA, Saenger W (1994) Hydrogen Bonding in Biological Structures, 2nd edn. Springer, Berlin Heidelberg

[5] Bernstein J, Etter MC, Leiserowitz L (1994) In: Bürgi H-B, Dunitz JD (eds) Structure Correlation, vol 2. VCH, Weinheim, p 431

[6] Lehn JM (1995) In: Supramolecular Chemistry: Concepts and Perspectives. VCH, Weinheim

[7] Lehn JM (1993) Science **260**: 1762

[8] Curtiss LA, Frurip DJ, Blander M (1979) J Chem Phys **71**: 2703

[9] Bürgi T, Droz T, Leutwyler S (1995) Chem Phys Lett **246**: 291

[10] Frey PA, Whitt SA, Tobin JB (1994) Science **264**: 1927

[11] Shan S, Loh S, Herschlag D (1996) Science **272:** 97

[12] Kato Y, Toledo LM, Rebek J Jr (1996) J Am Chem Soc **118**: 8575

[13] Schwarz B, Drueckhammer DG (1995) J Am Chem Soc **117:** 11902

[14] Ditchfield R (1976) J Chem Phys **65**: 3123

[15] McMichael Rohlfing C, Allen LC, Ditchfield R (1983) J Chem Phys **79**: 4958

[16] Berglund B, Vaughan RW (1980) J Chem Phys **73**: 2037

[17] Reimer JA, Vaughan RW (1980) J Magn Reson **41**: 483

[18] Gerald R, Bernhard T, Haeberlen U, Rendell J, Opella SJ (1993) J Am Chem Soc **115**: 777

[19] Wu CH, Ramamoorthy A, Gierasch LM, Opella SJ (1995) J Am Chem Soc **117**: 6148

[20] Ramamoorthy A, Wu CH, Opella SJ (1997) J Am Chem Soc **119**: 10479

[21] Llinás M, Wilson DM, Klein MP (1977) J Am Chem Soc **99**: 6846

[22] Llinás M, Horsley WJ, Klein MP (1976) J Am Chem Soc **98**: 7554

[23] Llinás M, Klein MP (1975) J Am Chem Soc **97**: 4731

[24] Facelli JC, Pugmire RJ, Grant DM (1996) J Am Chem Soc **118**: 5488

[25] Smirnov SN, Golubev NS, Denisov GS, Benedict H, Schah-Mohammedi P, Limbach HH (1996) J Am Chem Soc **118**: 4094

[26] Ash EL, Sudmeier JL, De Fabo EC, Bachovchin WW (1997) Science **278**: 1128

[27] Fujiwara FY, Martin JS (1974) J Am Chem Soc **96**: 7625

[28] Juranic N, Likic VA, Prendergas FG, Macura S (1996), J Am Chem Soc **118**: 7859
[29] Abragam A (1986) Principles of Nuclear Magnetism, Oxford Univ Press, Oxford
[30] Schmidt-Rohr K, Spiess HW (1994) Multidimensional Solid-State NMR and Polymers, Academic Press, London
[31] Shimizu H (1964) J Chem Phys **40**: 3357
[32] Mackor EL, MacLean C (1966) J Chem Phys **44**: 64
[33] Goldman M (1984) J Magn Reson **60**: 437
[34] Guéron M, Leroy JL, Griffey RH (1983) J Am Chem Soc **105**: 7262
[35] Burghardt I, Konrat R, Bodenhausen G (1992) Mol Phys **75**: 467
[36] Dalvit C, Bodenhausen G (1989) Chem Phys Lett **161**: 554
[37] Konrat R, Nutz K, Kalcher J, Sterk H (1994) J Phys Chem **98**: 7488
[38] McConnell HM (1956) J Chem Phys **25**: 709
[39] Konrat R, Sterk H (1993) Chem Phys Lett **203**: 75
[40] Reif B, Hennig M, Griesinger C (1997) Science **276**: 1230
[41] Yang D, Konrat R, Kay LE (1997) J Am Chem Soc **119**: 11938
[42] Tjandra N, Bax A (1997) J Am Chem Soc **119**: 8076
[43] Tessari M, Mulder FAA, Boelens R, Vuister GW (1997) J Magn Reson **127**: 128
[44] Tessari M, Vis H, Boelens R, Kaptein R, Vuister GW (1997) J Am Chem Soc **119**: 8985
[45] Tollinger M, Konrat R, Kräutler B (1997) J Mol Catal **116**: 147
[46] Konrat R, Tollinger M, Kräutler B (1998) In: Kräutler B, Arignoi D, Golding BT (eds) Vitamin B_{12} and B_{12}-Proteins. Wiley-VCH, Weinheim
[47] Tollinger M, Konrat R, Kräutler B (manuscript in preparation)
[48] Rossi M, Glusker JP, Randaccio L, Summers MF, Toscano PJ, Marzilli LG (1985) J Am Chem Soc **107**: 1729
[49] Liepinsh E, Otting G, Wüthrich K (1992) J Biomol NMR **2**: 447
[50] Leroy JL, Broseta D, Gueron M (1985) J Mol Biol **184**: 165
[51] Poppe L, van Halbeek H (1991) J Am Chem Soc **113**: 363
[52] (a) Carr HY, Purcell EM (1954) Phys Rev **94**: 630; (b) Meiboom S, Gill G (1958) Rev Sci Instrum **29**: 688
[53] Emsley L, Bodenhausen G (1989) J Magn Reson **82**: 211
[54] Emsley L, Bodenhausen G (1990) Chem Phys Lett **165**: 469
[55] Rhim WK, Burum DP, Elleman DD (1979) J Chem Phys **71**: 3139
[56] Ryan LM, Wilson RC, Gerstein BC (1977) Chem Phys Lett **52**: 341
[57] Pines A, Ruben DJ, Vega S, Mehring M (1976) Phys Rev Lett **36**: 110
[58] Hinton JF, Bennett DL (1985) Chem Phys Lett **116**: 292
[59] Schowen KB, Schowen RL (1982) Meth Enzym **87**: 551
[60] Hvidt A, Nielsen SO (1966) Adv Prot Chem **21**: 287
[61] Kresge AJ (1964) Pure Appl Chem **8**: 243
[62] Jarret RM, Saunders M (1985) J Am Chem Soc **107**: 2648
[63] Jarret RM, Saunders M (1986) J Am Chem Soc **108**: 7549
[64] Cleland WW (1980) Meth Enzymol **64**: 104
[65] Krevoy MM, Liang TM (1980) J Am Chem Soc **102**: 3315
[66] Cleland WW (1992) Biochemistry **31**: 317
[67] Weiss PM, Cook PF, Hermes JD, Cleland WW (1987) Biochemistry **26**: 7378
[68] Weiss PM, Boerner RJ, Cleland WW (1987) J Am Chem Soc **109**: 7201
[69] Loh SN, Markley JL (1993) In: Angletti R (ed) Techniques in Protein Chemistry IV. Academic Press, San Diego, p 517
[70] Loh SN, Markley JL (1994) Biochemistry **33**: 1029
[71] LiWang AC, Bax A (1996) J Am Chem Soc **118**: 12864
[72] Bothner-By AA, Stephens RL, Warren CD, Jeanloz RW (1985) J Am Chem Soc **106**: 811
[73] Piotto M, Saudek V, Sklenár V (1992) J Biomol NMR **2**: 661

[74] Van de Ven FJM, Hansen HGJM, Gräslund A, Hilbers CW (1988) J Magn Reson **79**: 221
[75] Davis DG, Bax A (1985) J Am Chem Soc **107**: 2820
[76] Englander SW, Kallenbach NR (1984) Quart Rev Biophys **16**: 521
[77] Bai Y, Milne JS, Mayne L, Englander SW (1993) Proteins **17**: 75
[78] Connelly GP, Bai Y, Jeng MF, Englander SW (1993) Proteins **17**: 87
[79] Englander JJ, Rogero JR, Englander SW (1985) Anal Biochem **147**: 234
[80] Eigen M (1964) Angew Chem **3**: 1
[81] Haslam JL, Eyring EM (1967) J Phys Chem **71**: 4470
[82] Rose MC, Stuehr J (1968) J Am Chem Soc **90**: 7205
[83] Linderstrøm-Lang KU (1955) In: Symposium on Peptide Chemistry. Chem Soc Spec Publ **2**: 1
[84] Linderstrøm-Lang KU (1958) In: Neuberger A (ed) Symposium on Protein Structure. Methuen London
[85] Woodward CK, Simon I, Tuchsen E (1982) Mol Cell Biochem **48**: 135
[86] Berger A, Linderstrøm-Lang KU (1957) Archs Biochem Biophys **69**: 106
[87] Wagner G (1983) Quart Rev Biophys **16**: 1
[88] Barksdale AD, Rosenberg A (1976) Meth Biochem Anal **28**: 1
[89] Zhang YP, Lewis RN, Henry GD, Sykes BD, Hodges RS, McElhaney RN (1995) Biochemistry **34**: 2348
[90] De Jongh HH, Goormaghtigh E, Ruysschaert JM (1995) Biochemistry **34**: 172
[91] Hildebrandt P, Vanhecke F, Heibel G, Mauk AG (1993) Biochemistry **32**: 14158
[92] Englander JJ, Calhoun DB, Englander SW (1979) Anal Biochem **92**: 517
[93] Kossiakoff AA (1982) Nature **296**: 713
[94] Pan Y, Briggs MS (1992) Biochemistry **31**: 11405
[95] Loh SN, Prehoda KE, Wang J, Markley JL (1993) Biochemistry **32**: 11022
[96] Rohl CA, Baldwin RL (1994) Biochemistry **33**: 7760
[97] Feng Y, Sligar FG, Wand AJ (1994) Nat Struct Biol **1**: 30
[98] Jeng M, Englander SW, Pardue K, Rogalskyi JS, McLendon G (1994) Nat Struct Biol **1**: 234
[99] Mayne L, Paterson Y, Cerasoli D, Englander SW (1992) Biochemistry **31**: 10678
[100] Ehrhardt MR, Urbauer JL, Wand AJ (1995) Biochemistry **34**: 2731
[101] Paterson Y, Englander SW, Roder H (1990) Science **249**: 755
[102] Forsén S, Hoffman RA (1963) J Chem Phys **40**: 1189
[103] Waelder S, Redfield AG (1977) Biopolymers **16**: 623
[104] Rosevaer PR, Fry DC, Mildvan AS (1985) J Magn Reson **61**: 102
[105] Krishna NR, Huang DH, Glickson JD, Rowan R, Walter R (1979) Biophys J **26**: 345
[106] Spera S, Bax A (1991) J Biomol NMR **1**: 155
[107] Sandström J (1982) In: Dynamic NMR Spectroscopy. Academic Press, New York
[108] Kratky C, Färber G, Gruber K, Wilson K, Dauter Z, Nolting HF, Konrat R, Kräutler B (1995) J Am Chem Soc **117**: 4654
[109] Bolton PH, Kearns DR (1979) J Am Chem Soc **101**: 479
[110] Pyle AM, Murphy FL, Cech TR (1992) Nature **358**: 123
[111] Fersht A (1985) In: Enzyme Structure and Mechanism. Freeman, New York
[112] Alber T, Sun DP, Wilson K, Wozniak JA, Cook SP, Matthews BW (1987) Nature **330**: 41
[113] Jucker FM, Heus HA, Yip PF, Moors EHM, Pardi A (1996) J Mol Biol **264**: 968

Received November 13, 1998. Accepted November 30, 1998

Mannich Bases as Model Compounds for Intramolecular Hydrogen Bonding II [1]. Structure and Properties in Solution

Aleksander Koll[1] and **Peter Wolschann**[2,*]

[1] Faculty of Chemistry, University of Wroclaw, PL-50383, Poland
[2] Institut für Theoretische Chemie und Strahlenchemie, Universität Wien, A-1090 Austria

Summary. Studies on hydrogen bond properties and on the structure of *ortho*-alkylaminomethylphenols and -naphthols in solution are presented. The main advantages of these systems are their stability and the defined stoichiometry of the molecules in comparison to intermolecularly hydrogen bonded complexes. Some specific properties of intramolecular hydrogen bonds are additionally discussed.

Keywords. *Mannich* bases; Alkylaminomethylphenol; Alkylaminomethylnaphthol; Intramolecular hydrogen bond; Proton transfer.

***Mannich*-Basen als Modellverbindungen für intramolekulare Wasserstoffbrückenbindungen, 2. Mitt. [1]. Struktur und Eigenschaften in Lösung**

Zusammenfassung. Eigenschaften und Strukturen der Wasserstoffbrücken von *ortho*-Alkylaminomethylphenolen und -naphtholen in Lösung werden präsentiert. Der große Vorteil dieser Verbindungen als Modelle für Wasserstoffbrückensysteme ist ihre große Stabilität und die Stöchiometrie dieser Verbindungsklasse in Vergleich mit intermolekularen Wasserstoffbrückenkomplexen. Einige spezielle Eigenschaften von intramolekularen Wasserstoffbrücken werden zusätzlich beschrieben und diskutiert.

Introduction

Hydrogen bonding between phenols and amines is of general interest because such systems are very convenient models for the investigation of the influence of various factors on hydrogen bond properties and proton transfer reactions in biomolecules. As in intermolecularly bonded complexes the stoichiometry may be uncertain and additional association steps can render the interpretation difficult, alternative model systems were selected. In particular, *ortho-Mannich* bases, the condensation

---

* Corresponding author

products of phenols or naphthols with aldehydes and secondary amines, seem to be very useful. The main advantage of these systems is the pronounced thermodynamic stability of the intramolecular hydrogen bonds. They permit the investigation of species characterized by the same stoichiometry and very similar structure in wide range of temperature and different surroundings without the need to determine equilibrium constants of a set of additional association steps and their variation under varying conditions as it is the case with intermolecular hydrogen bonding.

A large number of such intramolecularly hydrogen bonded compounds of the *Mannich* base type were studied in dependence of a variety of different parameters, like the properties of proton donor and acceptor components, temperature, environment, concentration, or phase. In the present paper we shall discuss the problems of structure and properties of *Mannich* bases upon variation of these parameters.

A specific feature of *ortho-Mannich* bases like that shown below is the bent hydrogen bond caused by steric circumstances, leading also to a nonplanar geometry of such molecules [1].

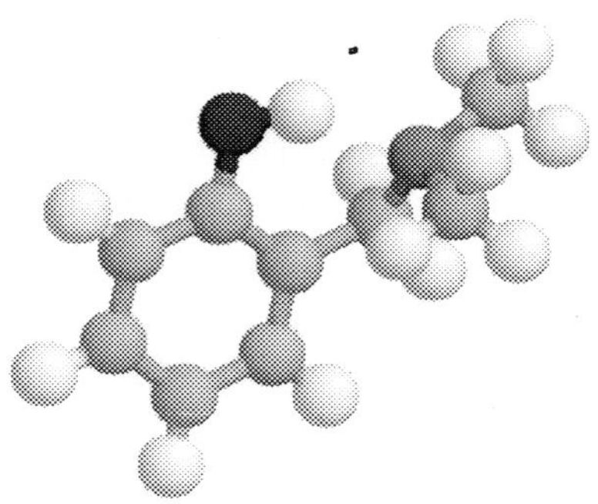

Discussion

Stoichiometry and stability of the intramolecular hydrogen bond in solution

One of the principal topics in hydrogen bond investigations is the stoichiometry of the considered complexes. In intramolecular hydrogen bonds, the situation seems to be simpler, because the stoichiometry is known. The limits of the stability of intramolecular hydrogen bonds as a function of concentration, solvents, and temperature has been of interest from the beginning of investigations on *Mannich* bases. It has been shown that in nonpolar and medium polar solvents (up to a permittivity of 37) the free ν_s(OH) bands do not appear in the IR spectra [2–6]. Moreover, strong dilution up to conditions under which analogous complexes of phenols with amines do not exist any more does not evoke appearance of such bands.

The results of IR absorption studies in the gas phase demonstrate the stability of the intramolecular hydrogen bond in *ortho-Mannich* bases even up to +150°C and a partial pressure of about 1 torr [7] (Fig. 1). Similar conclusions based on the results of dielectric measurements in solutions are presented in Refs. [8–11].

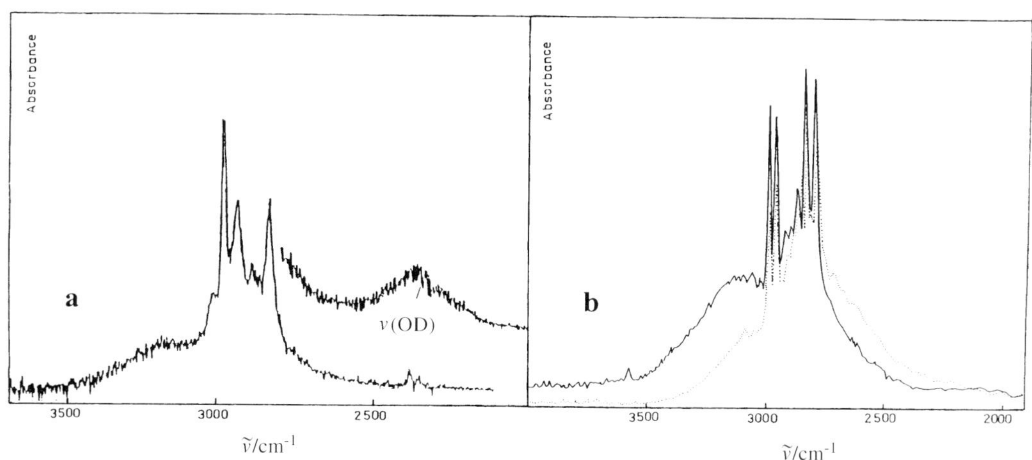

Fig. 1. IR spectra of a) 2-(N,N-diethylaminomethyl)-4,6-dimethylphenol in the gas phase (150°C, partial pressure 1 torr) and b) 2-(N,N-dimethylaminomethyl)-4,6-dichlorophenol in the gas phase (full line) and in CCl_4 solution (broken line, $c = 0.005\,mol \cdot dm^{-3}$ $d = 2\,cm$); Specord M-30 (Carl Zeiss Jena), resolution: $1.5\,cm^{-1}$

Structure of Mannich bases in solution

In the solid state, intramolecularly hydrogen bonded molecules or zwitterions forming cyclic dimers or chains were observed [1], depending on the difference of the acidity constants of the free proton donor and the protonated secondary amine (ΔpK_a), a parameter commonly used to characterize hydrogen bonds [12, 13]. Therefore, the question arises to what extent inter- and intramolecularly bonded species exist in solution in dependence on solvent polarity and temperature. It can be expected that nonionic structures with intramolecular hydrogen bonds as found in the solid state for *Mannich* bases with $\Delta pK_a < 2.5$ [1] should predominate in nonpolar solutions. Indirect support of such a suggestion is the similarity of the IR spectra of these compounds in CCl_4 solutions and in the solid state [3]. It was found, however, that for compounds with ΔpK_a above 2.5 the spectra may be pronouncedly different in both phases. The solution spectra in these cases resemble the spectra of molecular forms of hydrogen bonds in the solid state. This suggests that some *Mannich* bases which are of zwitterionic type in the solid state form intramolecular hydrogen bonds of molecular type in low polar solvents (Fig. 2) [14].

Systematic studies on a wide class of 75 different *Mannich* bases have shown that in CCl_4 solutions a drastic change of the IR spectra in the $\nu(OH)$ band region occurs for systems with $\Delta pK_a \approx 3.5$ [3]. At this value zwitterionic forms begin to appear. The position of the maximum of the absorption shifts in the direction characteristic for the ionic forms of hydrogen bonds (see Fig. 2). Intermediate frequencies are observed for systems exhibiting equilibria of molecular and ionic forms. Both OH and NH^+ bands of hydrogen bonded species are too broad to allow the discrimination of separate bands of particular forms.

Some direct information about the structure of intramolecular hydrogen bonds in solution of these compounds can be gained from dipole moment studies. For

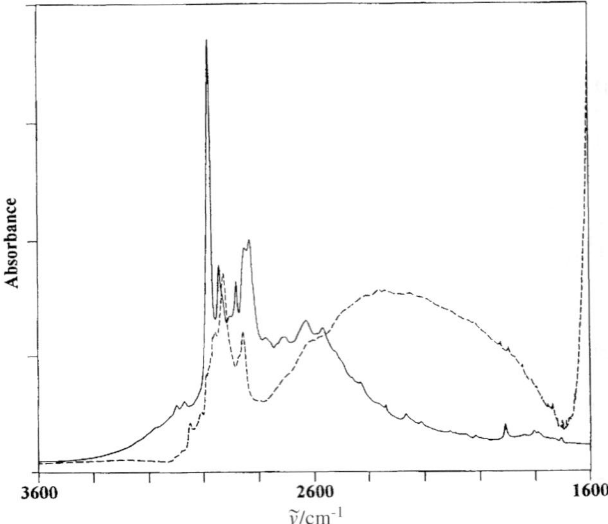

Fig. 2. Comparison of IR spectra in the ν(OH) absorption region of 2-(N,N-diethylaminomethyl)-4-nitrophenol in the solid state (KBr pellets, broken line) and in CCl_4 solution (full line, $c = 0.005 \, \text{mol} \cdot \text{dm}^{-3}$, $d = 9.55 \, \text{mm}$); Nicolet 205, resolution: $2 \, \text{cm}^{-1}$

intermolecular complexes, a sigmoidal dependence of the polarity of the hydrogen bond ($\Delta\mu$) on ΔpK_a (Fig. 3) was observed [15–18]. Calculated values for *Mannich* bases assuming nonplanar geometries (found for molecular hydrogen bonded structures in the solid state) show a very good coincidence with this dependence which was interpreted in terms of proton transfer equilibria [12, 15]. A small increase of μ in the order of 0.5–1.0 D is a consequence of mutual polarization of the complex components. A stronger increase of μ, as observed for *Mannich* bases, indicates the existence of some amounts of ionic forms in solution [19].

Formation of cyclic dimers, as observed for some solid states structures [1], would lead to a decrease of the dipole moment; the increase stated in Ref [19]

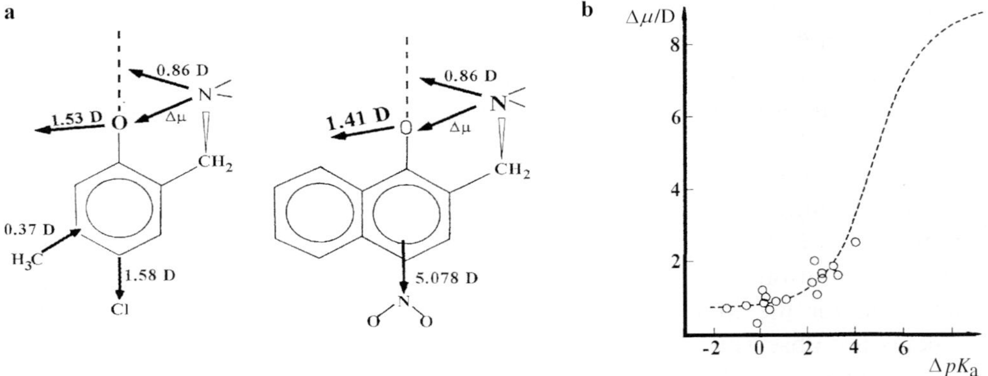

Fig. 3. a: Scheme of vector addition for some representative *Mannich* bases; b: Dependence of the hydrogen bond polarity ($\Delta\mu$) on ΔpK_a in benzene; open circles: data for *Mannich* bases [19], dotted line: intermolecular complexes [15]

suggests that intramolecular zwitterionic forms or chains are formed. The complexes with ΔpK_a below 4 do not show a concentration dependence of experimental dipole moments. From that it follows that in solutions of such *Mannich* bases no aggregation of the zwitterionic structures occurs. The influence of the solvents electric permittivity increase on dipole moments has been studied as well [20]. For a compound with a ΔpK_a of 2.72, no such dependence was found up to a permittivity of 10. Compounds with $\Delta pK_a = 3.04$ and 3.72 show an increase of experimental dipole moments with permittivity, thus demonstrating that the formed ionic species are not associated to cyclic dimers.

A model of proton transfer equilibria was verified by measurements of the temperature dependence of dipole moments in solutions [21], and full agreement of the obtained results with the model was found. A specific feature of these studies should be mentioned: due to the large stability of intramolecular hydrogen bonds, $\Delta\mu$ changes were caused by permittivity and temperature variation over a broad range for the particular compound rather than by changing the substituents. In this case, the mutual polarization of the complex components was also modified; it influences the shape of the $\Delta\mu$ vs. ΔpK_a dependence [15–18, 22].

Proton transfer in Mannich bases

From the point of view of polarity, the complexes with hydrogen bonds can be divided into three classes (A, B, and C) [18]. For weak hydrogen bonds, an increase of the dipole moment relative to the vector sum of the dipole moments of the components is less than 1 D (class A, Fig. 4).

At the other extreme, proton transfer may lead to a dipole moment increase in the order to 10 D [15] (class C). In many cases some intermediate increase of polarity is observed (class B). For such compounds, a strong dependence of the dipole moment of ΔpK_a (as well as on solvent or temperature) indicates a dynamic proton transfer equilibrium in solution [12, 15].

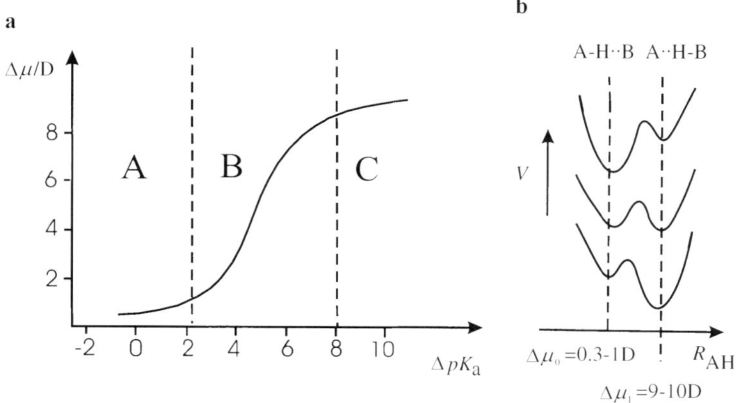

Fig. 4. a: Ranges of existence of A) molecular forms, B) proton transfer equilibria, C) ionic forms defined on the basis of the dependence of $\Delta\mu$ on ΔpK_a; b: schematic profile of potential energy for the proton movement within the hydrogen bridge for the cases A, B, and C

According to the above studies, *Mannich* bases with intramolecular hydrogen bonds belong to class A in nonpolar solvents and some derivatives with ΔpK_a values above 3 can be assigned to class B. By variation of temperature or/and solvent polarity the proton transfer equilibrium can be shifted, thus modifying the structure of the studied species in solution.

The ^{1}H NMR chemical shift dependence on ΔpK_a observed for OH protons in *Mannich* bases [23] has a shape typical for systems with proton transfer equilibria. A strong influence of the surroundings on the position of the maximal chemical shifts has been stressed. Temperature decrease leads to an enhancement of effects characteristic for proton transfer.

Clear indications of proton transfer equilibria in *Mannich* bases can be found in the electronic absorption spectra of these compounds. As electronic transitions are much faster (10^{-15} s) than any chemical reaction (even extremely fast intra-molecular proton transfers), it can be easily distinguished between the molecular and the ionic forms of the hydrogen bond. The electronic absorption bands of phenols and phenolates as well as naphthols and naptholates are different [24, 25]. As an example, the spectra of 2-morpholinomethyl-naphthol-1 in different solvents are shown in Fig. 5. The spectral differences are caused by the shift of the proton transfer equilibria with variation of solvent polarity.

In some *Mannich* bases double fluorescence was observed which could be attributed to emission from the molecular and zwitterionic type of molecules [26]. Double fluorescence is also observed for molecules with weak hydrogen bonds *e.g.* in 6-(N,N-diethylaminomethyl)-2,3,5-trimethylphenol ($\Delta pK_a = -1.04$) [27], which does not exist in the zwitterionic form in the ground state. In the singlet excited state the proton transfer reaction proceeds to a greater extent than in the ground state, since pK_a values of phenols and naphthols may increase over six units in the excited state [26].

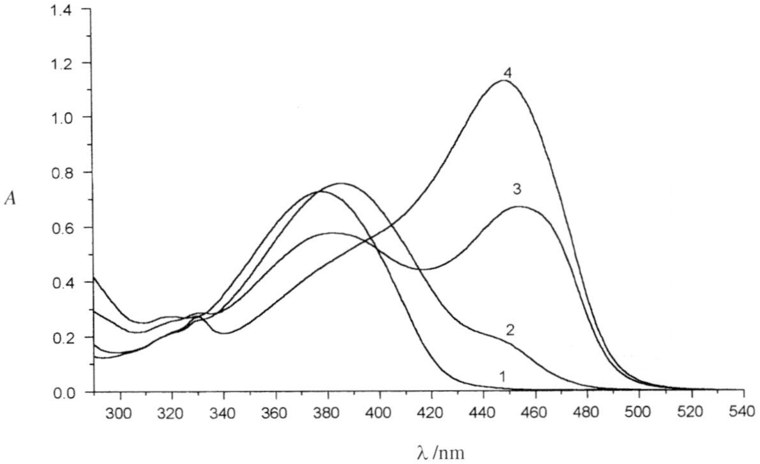

Fig. 5. Influence of solvents on UV/Vis spectra of 2-morpholinomethyl-naphthol-1 ($c = 4 \cdot 10^{-4}$ mol$\cdot$dm^{-3} 25°C, Hitachi U-3501); 1: dioxane, 2: acetonitrile, 3: ethanol, 4: ethanol-water (10%)

It appears that also in the excited state the stabilization of the zwitterionic forms by the solvent is a necessary condition to observe the proton transfer states. In the case of the above mentioned molecule an increase of electric permittivity above 6 is a prerequisite to observe double fluorescence. Such a solvent interaction has been considered on the basis of the dielectric continuum model [28].

Proton transfer equilibrium constants and related thermodynamic parameters

Using electron absorption spectra, the amount of both forms existing in a chemical equilibrium as well as the equilibrium constant can be determined precisely. The thermodynamic characteristics of the proton transfer reaction can also be studied when temperature dependent measurements are performed [29]. Such studies are described in Refs. [4, 29, 30], and examples of estimated thermodynamic parameters in the proton transfer process [30] are presented in Table 1.

For all cases where proton transfer equilibria were detected it was found that $\Delta H°$ of the proton transfer reaction is negative within the range from -6 to -17 kJ/mol [29]. In agreement with negative values of $\Delta H°_{PT}$, a shift of the equilibrium towards the ionic forms was observed with decreasing temperature [25]. The IR spectra also allow a direct determination of the proton transfer equilibrium position [4]. The band at $1185\,cm^{-1}$ of 6-diethylaminomethyl-2,3,4,5-tetrachlorophenol, which has $\gamma(O–H)$ bending and $\nu(C–O)$ stretching character, disappears after deuteration, and a new band at $1025\,cm^{-1}$ appears. The intensity of the $1185\,cm^{-1}$ band was used as a measure of the amount of the molecular form. The degree of proton transfer was determined in solvents with electric permittivity ranging from 2 to 38 in dependence of temperature [4]. Linear correlations between $\log K_{PT}$ (the proton transfer equilibrium constant) and *Onsager* parameters as well as the solvent polarity parameter E_T were observed. It was shown that enhancement of the solvent polarity shifts the equilibrium in the proton transfer direction. Thermodynamic characteristics of the proton transfer reaction obtained by IR spectroscopy [4] agree well with the results of more precise UV/Vis techniques [29]. $\Delta S°_{PT}$ values are negative (within the range of -24 to -62 J/mol $\cdot$ deg) with rather high absolute values as for internal proton transfer only [31], suggesting solvent

Table 1. Thermodynamic parameters of the proton transfer reaction in *Mannich* bases from derivatives of 4-nitrophenol (**1**) and 2,4,5-trichlorophenol (**2**) with benzylamines in methanol [30]

R	pK_a^a	$\nu/10^2\,Å^3$	$-\Delta H°_{PT}/kJ\cdot mol^{-1}$		$\Delta G°_{PT}/kJ\cdot mol^{-1}$		$-\Delta S°_{PT}/J\cdot mol^{-1}K^{-1}$	
			1	**2**	**1**	**2**	**1**	**2**
CH_3	8.91	3.66	9.20	10.60	0.083	0.498	31.64	36.80
C_2H_5	9.44	2.09	8.74	9.42	2.002	0.260	36.04	32.49
C_3H_7	9.40	1.47	8.32	8.62	2.40	0.54	35.96	30.72
C_4H_9	9.41	1.13		8.05		0.76		29.54
C_6H_{13}	9.50	0.77	7.10	7.51	3.44	1.17	35.46	29.15
$C_{10}H_{21}$	9.45	0.47	5.89	5.65	3.81	1.63	32.69	24.43

[a] pK_a value of the corresponding benzylamine; volume of the alkyl residue

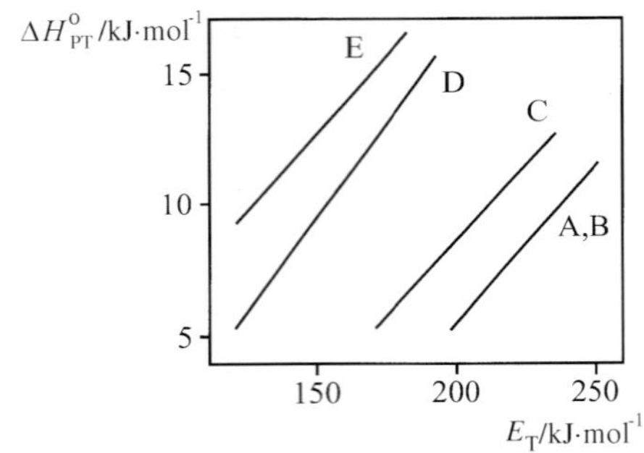

Fig. 6. Correlations between $\Delta H^{\circ}_{\mathrm{PT}}$ and the solvent polarity parameter E_{T} [30] for A: 2-(N,N-dimethylaminomethyl)-4,6-dibromophenol ($\Delta pK_{\mathrm{a}} = 1.1$), B: 2-(N,N-dimethylaminomethyl)-4-nitrophenol ($\Delta pK_{\mathrm{a}} = 1.8$), C: 2-(N,N-diethylaminomethyl)-3,4,6-trichlorophenol ($\Delta pK_{\mathrm{a}} = 2.7$), D: 2-(N,N-diethylaminomethyl)-3,4,5,-tetrachlorophenol ($\Delta pK_{\mathrm{a}} = 3.0$), E: 2-(N,N-diethylaminomethyl)-4-nitronaphthol-1 ($\Delta pK_{\mathrm{a}} = 3.7$)

participation in the proton transfer reaction. The solvent polarity increase generally leads to an enhancement of the absolute values of $\Delta H^{\circ}_{\mathrm{PT}}$ and $\Delta S^{\circ}_{\mathrm{PT}}$. A linear dependence of $\Delta H^{\circ}_{\mathrm{PT}}$ on solvent polarity (E_{T}) with more or less the same slope but distinctly different intercepts for various *Mannich* bases was found [29, 30] (Fig. 6).

A linear correlation between $\Delta H^{\circ}_{\mathrm{PT}}$ and ΔpK_{a} was found as well [30]. The results show that there are two effects responsible for the existence and the position of the proton transfer equilibrium. The first might be called internal (see also Ref. [32]) – it depends on the relation between the electronic energy of hydrogen bonded and proton transferred forms in the gas phase. It seems that these energy levels are shifted always in favour of the molecular forms; ionic forms of *ortho*-*Mannich* bases in the gas phase were not detected. Theoretical calculations also yield a lower energy for the molecular forms [33]. The reason that the proton transfer forms are observed in solution is the external interaction with the solvent molecules. For weak hydrogen bonds, the electronic level distribution is so unfavourable for the proton transfer forms that even the strong interaction with a solvent cannot force the proton transfer to occur. The electronic absorption spectra of 2-(N,N′-dimethylaminomethyl)-4-chloro-5-methylphenol ($\Delta pK_{\mathrm{a}} = -0.6$) in butanol-1($\varepsilon = 17.8$) at its freezing temperature (110 K) did not give any evidence for the absorption of proton transfer forms [34]. Some amount of proton transfer forms in this compound was found, however, in glycerol ($\varepsilon = 45.8$) at temperatures lower than 220 K.

$\Delta G^{\circ}_{\mathrm{PT}}$ values for all studied cases were found to be small and positive at room temperature (within a range from $+0.1$ to 8.5 kJ/mol [29]). The relatively large negative entropy of the proton transfer reaction decreases the amount of proton transfer forms. In Ref. [35], however, all numerical values of the thermodynamic characteristics of the proton transfer process have to be considered as overestimated, because the change of electric permittivity with temperature was not accounted for.

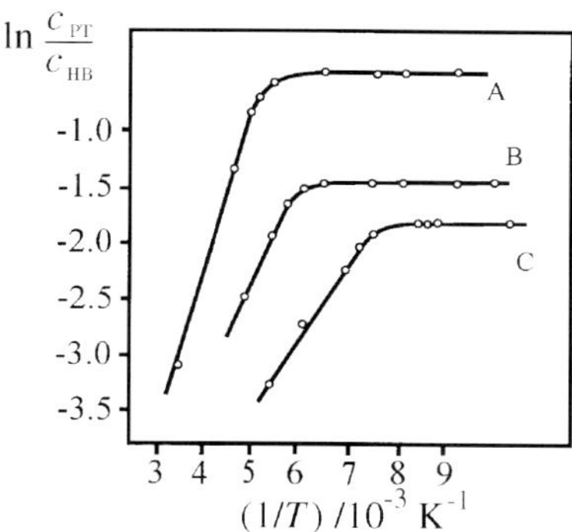

Fig. 7. Dependence of the logarithm of the concentration ratio of zwitterionic (PT) and molecular (HB) forms on $1/T$; A: 2-(N,N-dimethylaminomethyl)-4-nitrophenol in squalane, B: 2-(N,N-dimethylaminomethyl)-tetrachlorophenol in squalane, C: 2-(N,N-dimethylaminomethyl)-tetrachlorophenol in cumol

Influence of the solvent on proton transfer equilibrium

A very strong support for the hypothesis of solvent participation in proton transfer reactions arises from studies at low temperatures [34]. Crystallization of the solution prohibits the reorientation of the solvent and thus freezes the equilibrium. Similar observations were made by FTIR spectroscopy, and related conclusions were drawn [36].

The importance of solute-solvent interaction was also demonstrated in a study on the influence of steric effects on the proton transfer equilibrium of 4-nitrophenol and 2,4,5-trichlorophenol derivatives in methanol [30]. In condensation products of these phenols with secondary amines, a gradual decrease of the amount of proton transfer forms was found with increasing N-chain length (cf. Tab. 1). It was concluded that bulky substituents shield the center of the intramolecular proton transfer reaction from the polar solvent molecules. The effect was found to be inversely proportional to the molecular volume of the substituents [30], as it follows from the *Onsager* reaction field theory:

$$\Delta G_{PT}^{0(s)} = \Delta G_{PT}^{0(g)} - N_A \frac{\mu_{PT}^2 - \mu_{HB}^2}{a^3} \frac{\varepsilon - 1}{2\varepsilon + 1} \tag{1}$$

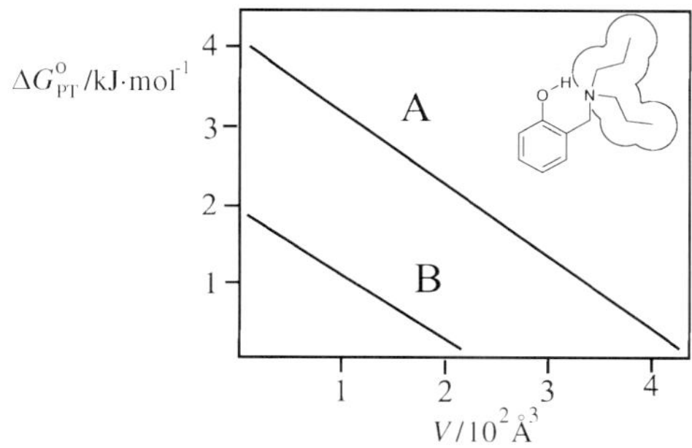

Fig. 8. Correlation between ΔG_{PT}° and V of aliphatic chains for 2-(N,N-dialkylaminomethyl)-4-nitrophenols (A) and -3,4,6-trichlorophenols (B) in methanol

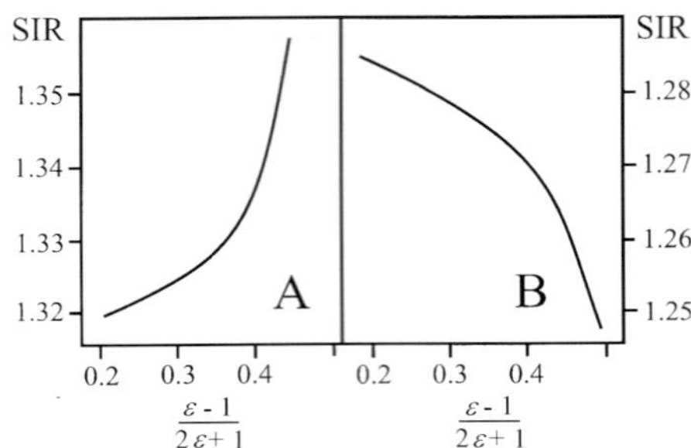

Fig. 9. Dependence of the spectroscopic isotopic ratio (SIR; A: $\nu^+_{\mathrm{NH}}/\nu^-_{\mathrm{NH}}$, B: $\nu_{\mathrm{OH}}/\nu_{\mathrm{OD}}$) on the *Onsager* parameter

Another example of the influence of the environment on the proton transfer equilibrium was demonstrated in Ref. [37]. Piperidinomethyl-2-naphthols form inclusion complexes with β-cyclodextrins in aqueous solution. The equilibrium constant of the proton transfer from the neutral to the zwitterionic form is strongly reduced by complexation into the cyclodextrin cavity, which is much less polar than the solvent.

In systems where the shift of the equilibrium towards one species – molecular or zwitterionic form – is complete, one can follow the influence of solvent and of temperature on the properties of particular forms of the hydrogen bond. For example, a comparison of the IR spectra of the molecular form of *Mannich* bases in the gas phase and in CCl_4 solution has been performed [7] (cf. Fig. 1). It was shown that the transfer from the gas phase to solution results in a shift of the $\nu(\mathrm{OH})$ frequency to lower values by 100–150 cm^{-1}; larger shifts correspond to molecules with stronger hydrogen bonds. On the basis of a correlation of $\Delta H°$ of the hydrogen bond formation with $\Delta\nu(\mathrm{OH})$, an increase of the intramolecular hydrogen bond strength over an order of at least 4 kJ/mol was found upon passing from the gas phase to a nonactive solvent like CCl_4 [7].

A very interesting dependence of the spectroscopic isotopic ratio (SIR) on solvent polarity has been established [37]. For molecular hydrogen bonds the SIR value decreases with the solvent polarity, suggesting a strengthening of the hydrogen bond. The opposite effect was found in the case of hydrogen bonds in the zwitterionic form.

Considerations of the proton transfer potentials by IR spectroscopy

Extensive studies of the influence of the temperature on the $\nu(\mathrm{OH})$ band shape have been performed. In the gas phase, in liquified nobel gases, or in solid matrices of nobel gases these bands are very broad (400–600 cm^{-1} halfwidth) [7, 39–41]. A very strong broadening was found particularly in 2,6-*bis*-(diethylaminomethyl)-phenols where collective proton motion can be anticipated [36, 42, 43]. A very broad absorption was also observed in monosubstituted *Mannich* bases in solvents of higher polarity, *e.g.* in acetonitrile [4]. The broadening was explained by a proton fluctuation mechanism and strong interactions with the environment [44]. In the gas phase and in nobel gas matrices the broadening can be explained only by a coupling of the $\nu(\mathrm{OH})$ mode with internal modes of the molecule. The low frequency bending vibrations of the chelate ring, modulating the N$\cdots$O distance, were identified as those internal modes which are coupled with the $\nu_s(OH)$ vibration on the basis of semiempirical force field calculations and experimental FTIR spectra [40, 41]. The band shape of the $\nu(OH)$ absorption could be reproduced on the basis of this model. Weaker coupling in OD samples than in nondeuterated ones was mentioned. The temperature dependence of spectral moments, also for deuterated samples, was studied in a wide temperature range (170–380 K).

Specific properties of the hydrogen bond in Mannich bases

When discussing different aspects of the hydrogen bond on the example of *ortho-Mannich* bases, the question arises to what extent the observed properties of the hydrogen bond are of general importance, *i.e.* what are the specific features of the intramolecular hydrogen bond in *Mannich* bases.

Comparative studies of intermolecular complexes of phenols with amines and the *Mannich* bases synthesized from these components have been performed [3]. For bent intramolecular systems the $\nu_s(OH)$ absorption maximum was shifted to higher frequencies by less than $100\,\mathrm{cm}^{-1}$ relative to analogous intermolecular systems (Fig. 10) [3].

Taking into account the weakening of the base strength in *Mannich* bases (pK_a values of benzylamines are lower than those of analogous aliphatic amines [45]), it can be concluded that the properties of both types of hydrogen bonds are very similar. The identity of the correlations between polarity ($\Delta\mu$) and ΔpK_a for *Mannich* bases and intermolecular complexes has been mentioned above.

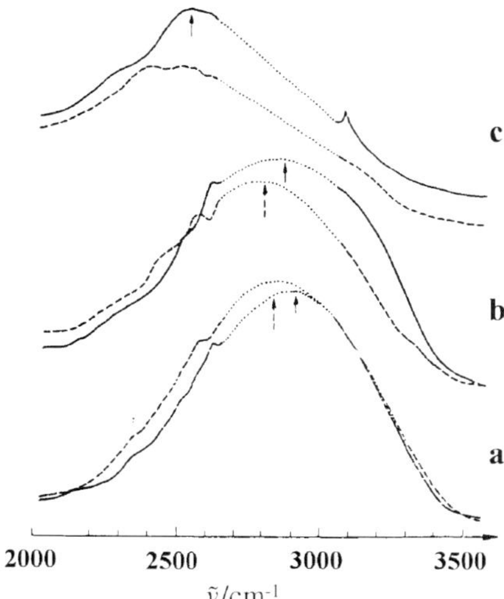

Fig. 10. Comparison of $\nu(OH)$ of *Mannich* bases (full lines, $c = 0.001\,\mathrm{mol\cdot dm^{-3}}$, $d = 1\text{-}4\,\mathrm{cm}$) and complexes of related phenols and amines (dotted lines, $c = 0.01\,\mathrm{mol\cdot dm^{-3}}$, $d = 1\text{–}2.6\,\mathrm{mm}$) in CCl$_4$ (Perkin-Elmer 621) [3]; a: 2-(N,N-diethylaminomethyl)-2,4-dimethylphenol/2,4-dimethylphenol+ triethylamine, b: 2-(N-piperidinomethyl)-4-chlorophenol/4-chlorophenol+N-methylpiperidine, c: 2-(N,N-diethylaminomethyl)-3,4,6-trichlorophenol/2,4,5-trichlorophenol+triethylamine

$$\frac{\partial \mu}{\partial r} \cong \frac{\partial \mu(OH)}{\partial r} + a \times \frac{\partial \mu(LP)}{\partial r}$$

$$\frac{\partial \mu}{\partial r} \cong \frac{\partial \mu(OH)}{\partial r} + \frac{\partial \mu(LP)}{\partial r} \times \cos\alpha$$

Scheme 1

The results of hydrogen bond studies in *ortho-Mannich* bases seem to be of general importance. Some peculiarities resulting from the hydrogen bridge bending in *Mannich* bases should be mentioned, however. They concern mainly the integrated intensity of ν(OH) bands which are related approximately to the square of the dipole moment derivative with respect to the O–H bond stretching (Scheme 1).

The O–H stretching vibrations lead to a polarization of both proton donor and acceptor molecules (mainly the lone pair of the latter); $\partial\mu/\partial r$ is a vector sum of both components. The second takes into consideration the α with the OH axis in *Mannich* bases, lead to a decrease of the total $\partial\mu/\partial r$ value. Such suggestions follow also from semiempirical calculations. An experimental verification of these considerations is shown in Fig. 11 [46].

It can be seen that the integrated intensity in bent hydrogen bonds is reduced in comparison to analogous intermolecular, more linear hydrogen bonds. The very good linear correlation between intensity and ΔpK_a should be mentioned. The point with increased intensity (second square from left side in Fig. 11) with respect to the correlation for *Mannich* bases belongs to 2-(N,N-dimethylaminomethyl)-3,4,5-trimethylphenol, where steric repulsion by the 3-methyl group leads to hydrogen bond shortening and, therefore, strengthening [47]. The integrated intensity for this compound appears above the correlation [46] which takes into account mainly acid-base properties of the system. In more extended studies [48] these results are compared with data for analogous *Schiff* bases with direct π-electronic coupling between acidic and basic centers. The intensity was found to decrease in

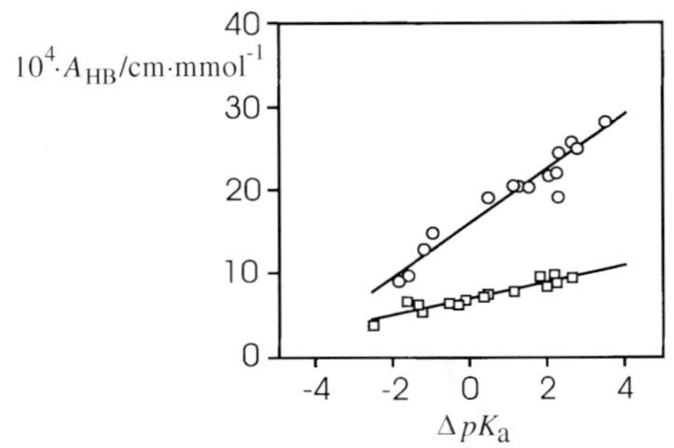

Fig. 11. Dependence of the integrated intensity of the ν(OH) band of *Mannich* bases (□) and complexes of N,N-dimethylamine with related phenols (o) on ΔpK_a

comparison to analogous *Mannich* bases, whereas the frequency shift $\Delta\nu(\text{OH})$ was larger. It can be concluded that particular hydrogen bonds of the O–H$\cdots$N type in *Mannich* bases, *Schiff* bases, or intermolecular complexes between phenols and amines have specific individual characteristics, *e.g.* different slopes and intercepts for $\Delta\nu(\text{OH})$ *vs.* ΔpK_a, intensity of $\nu(\text{OH})$ *vs.* ΔpK_a, or intensity *vs.* $\Delta\nu(\text{OH})$ correlations. Studies of such specific collective characteristics of particular groups of hydrogen bonded systems are important because they provide a better understanding of the nature of hydrogen bonding. Parameters of such characteristics can be verified by quantum mechanical calculations.

Self-association of polar forms of Mannich bases

Some limitations in the interpretation of the results for *Mannich* bases may result from the possible self-association of their polar forms. Thus, the position of the proton transfer equilibrium of 6-(N,N-diethylaminomethyl)-2,5-dinitrophenol ($\Delta pK_a = 4.04$) was found to depend on concentration [49]. It was concluded that some aggregation has to be considered for solutions of such *Mannich* bases. In order to verify this conclusion, measurements of the average molecular weight by the vapor pressure osmometric technique were performed [50]; the results are depicted in Fig. 12.

6-(N,N-diethylaminomethyl)-2,3,4,5-tetrachlorophenol ($\Delta pK_a = 3.04$) does not show any association over the whole concentration range (up to $4 \cdot 10^{-2}$ mol · dm^{-3}) in benzene solutions at room temperature. About 9% of the ionic form were found in benzene at room temperature [20]. An increase of the temperature within the range of 283–328 K leads to a decrease of the polarity of the hydrogen bond from 1.95 to 1.63 D caused by a decrease of the amount of ionic forms [21]. The dipole moment at room temperature increases from 4.9 to 5.6 D with increasing solvent polarity ($\varepsilon = 2$ to 10), suggesting an increase of the amount of ionic forms

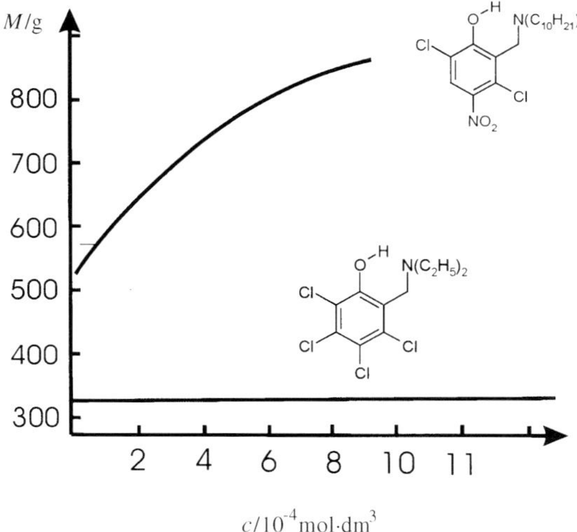

Fig. 12. Dependence of the average molecular weight (M) on the concentration of 6-(N,N-dipropylaminomethyl)-tetrachlorophenol and 6-(N,N-didecylaminomethyl)-2,5-dichloro-4-nitrophenol in benzene at room temperature

without formation of cyclic dimers. The amount of ionic forms seems to be too low to allow for self-association at these experimental conditions.

The ionization and association processes for this compound were also studied at low temperatures in $CDCl_3$ and CD_2Cl_2 *via* observation of the secondary isotope effect in ^{13}C NMR connected with D/H exchange of the OH group of *Mannich* bases. It was shown that the charge redistribution upon the proton transfer leads to an inversion of the sign of this effect [51].

In the case of 6-(N,N-diethylaminomethyl)-2,4,5,6-tetrachlorophenol, at 245 K in CH_2Cl_2 complete compensation of hydrogen bonding and proton transfer effects proceeds at about 20% contribution of the ionic form as established by US-Vis spectroscopy. Further decrease of temperature leads to an increase of the isotope effect in accordance with the increasing amount of ionic forms. At about 200 K, a surprisingly quick diminishing of these effects was observed. The authors related this to the self-association to cyclic dimers which causes a redistribution of atomic charges; the amount of ionic forms grows continuously till 178 K. In 6-(N,N-diethylaminomethyl)-2,4,5-trichlorophenol ($\Delta pK_a = 2.64$) no temperature effects were observed.

In the case of 6-(N,N-didecylaminomethyl)-2,5-dichloro-4-nitrophenol ($\Delta pK_a = 4.40$) at room temperature an enhanced molecular weight (about 10%) was detected even at such a low concentration as $1.10^{-3}\,mol\,dm^{-3}$ [50, 52] (Fig. 12). An increase of the concentration up to almost saturated solutions leads to molecular weights even higher than those expected for dimers. The dependence of dipole moments of this compound on concentration in benzene shows that the aggregates have a cyclic form; a decrease of the average dipole moment with increasing concentration was found. On the other hand it was stated that an increase of the concentration leads to an intensity enhancement of the electronic absorption in the region characteristic for proton transfer forms. This indicates that the proton transfer reaction is strongly coupled with the association process. A solvent polarity increase like in chloroform, dichloromethane, or dichloroethane leads to stronger association. This suggests that the solvent assisted proton transfer process is a primary effect. When the amount of the ionic forms is sufficiently large and the specific solvent interactions are not too high like, for example in alcohols, the aggregation process starts. Generally, this problems become important at room temperature only in low polar solvents and for molecules with potentially very high ability to ionization. Considering the results shown in Fig. 4 one can expect a boundary condition of $\Delta pK_a \approx 4$. At $\Delta pK_a = 4$ still an increased value of $\Delta\mu$ was observed (Fig. 3). Similarly, a radical change of $\nu(OH)$ for CCl_4 solutions of *Mannich* bases was observed at $\Delta pK_a \approx 4$ [3].

General description of the proton transfer equilibria in solutions

In many of the papers mentioned above, information concerning proton transfer equilibria, in particular those of *Mannich* bases, was presented. In this chapter we aim to present a general description of this equilibrium in order to establish the regions of its existence as well as regions of existence of separate forms in dependence on solvent polarity, ΔpK_a, and temperature. Such a description can be useful for the interpretation of experimental results and may be used in planning further experiments.

As literature shows [4, 30, 52] there exists a linear correlation between $\Delta H°$ of the proton transfer reaction and the solvent polarity parameter E_T with similar slopes for different *Mannich* bases, but shifted in relation each to other. It was also demonstrated that the distance between those correlation lines depends on ΔpK_a; a linear correlation was found for the relation between $\Delta H°_\mathrm{PT}$ and ΔpK_a in a given solvent.

A two parameter regression line was found describing the dependence of $\Delta H°_\mathrm{PT}$ on both parameters at room temperature:

$$-\Delta H°_\mathrm{PT} = 0.106E_\mathrm{T} + 9\Delta pK_\mathrm{a} - 36 \tag{2}$$

No temperature dependence of $\Delta H°_\mathrm{PT}$ was detected exceeding the range of the experimental error of its determination. Equation (2) reproduces the results fulfilling separate one-parameter correlation lines with a precision of $\pm 1\,\mathrm{kJ/mol}$.

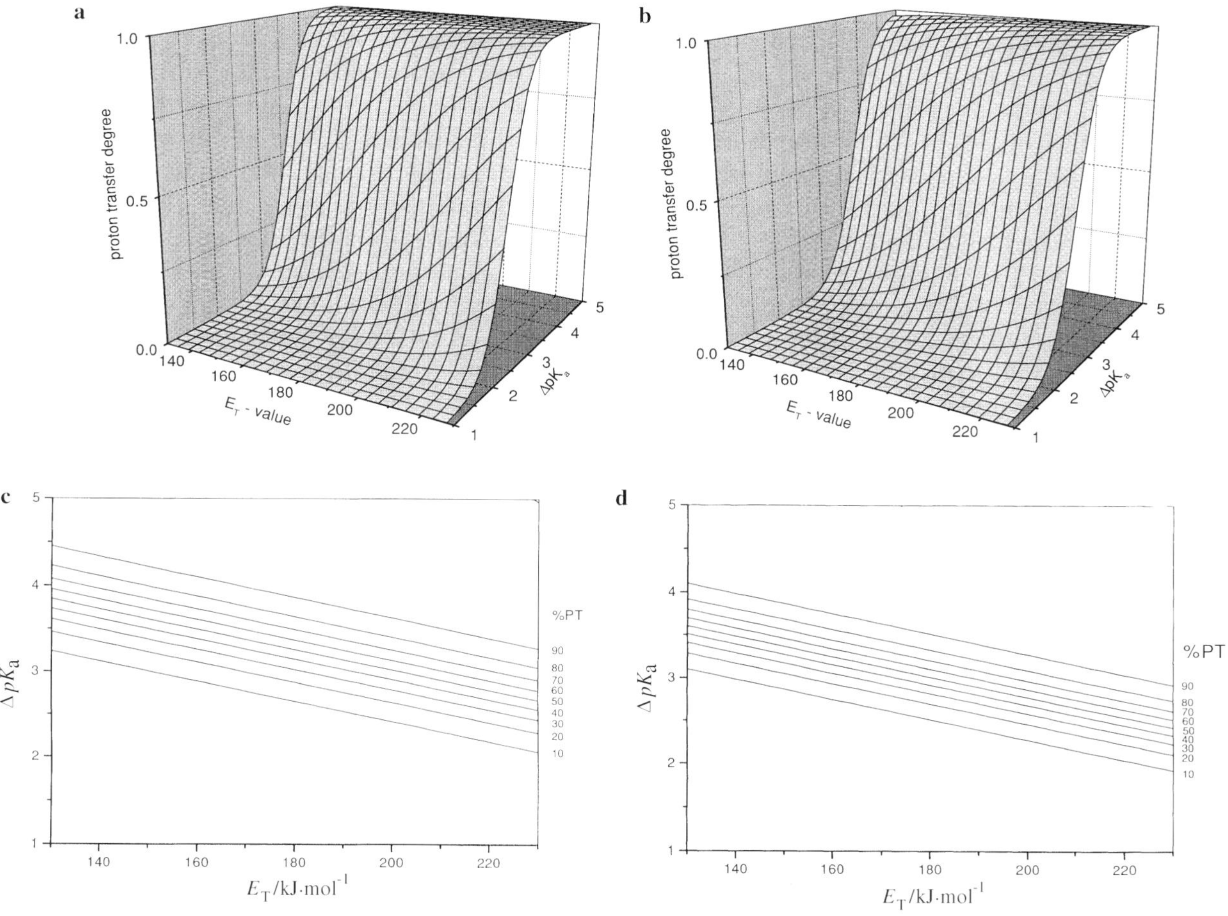

Fig. 13. Dependence of the proton transfer degree on E_T of solvents and ΔpK_a of *Mannich* bases;
a: 298 K, b: 200 K, c: 298 K, 2D contour plot, d: 200 K, 2D contour plot

To account for the temperature dependence of the equilibrium constant, the above equation was combined with the *van't Hoff* relation:

$$\ln K = (0.106 E_T + 9\Delta pK_a - 36) \cdot 1/(RT) + C \tag{3}$$

The parameter C was determined using the results of temperature dependence measurements of the proton transfer equilibrium in 6-(diethylaminomethyl)-2,3,4,5-tetrachlorophenol in a few solvents [4]. The C value was estimated as equal to -5 ± 0.17.

Equation (3) was verified on results not used for its calibration. For example, the amount of ionic species of 6-(dimethylaminomethyl)-2,3,4,5-tetrachlorophenol in CCl_4 at room temperature calculated from this equation is 6% which is in accordance with the results discussed in Refs. [20, 21, 51]. The IR results [3] show a pronounced increase of the $\Delta\nu(OH)$ values for *Mannich* bases in CCl_4 at room temperature at $\Delta pK_a\approx4$. Equation (3) affords about 68% of proton transfer at these conditions, which agrees with the results presented in Ref. [3]. The equation can be presented in a graphical way, which helps to understand the influence of such parameters as the solvent polarity (E_T), ΔpK_a, and temperature on the position of the proton transfer equilibrium. Figure 13 shows such plots for 298 and 200 K.

At room temperature, the range of existence of molecules with less than 10% of proton transfer extends to $\Delta pK_a = 3.2$ for low-polar solvents. This border shifts to 2.1 for polar solvents like alcohols. 50% of proton transfer is reached at ΔpK_a values of 3.9 at room temperature in nonpolar solvents and 2.7 in highly polar solvents. Decrease of the temperature to 200 K obviously increases the regions of proton transfer and decreases the region of existence of molecular forms. Nevertheless, a large variety of molecular forms is observed at low temperatures. Steeper dependences of the degree of proton transfer with respect to ΔpK_a are observed for low temperatures than for room temperature. Figure 13 allows to estimate the proton transfer equilibrium at each state described by particular values of E_T and ΔpK_a at temperatures of 298 and 200 K. Situations at intermediate temperatures can be analyzed using Eq. (3).

Conclusions

Results obtained from investigations of intramolecular hydrogen bonding in *Mannich* bases are presented and discussed in this review. *Mannich* bases have been used as model systems to study the nature of the hydrogen bond as well as the environmental and temperature influence on hydrogen bond properties including the proton transfer reaction. The advantages of these compounds are the high stability of the hydrogen bond, the defined stoichiometry, and the presence of similar structures in different phases and solvents, and at different temperatures. Additionally, direct electronic coupling between acidic and basic centres is substantially reduced by the methylene bridge. Very specific features of so-called resonance assisted hydrogen bonds do not influence the final conclusions resulting from the experimental studies.

Evidences of the high stability of these intramolecular hydrogen bonds are provided by IR and dipole moment studies. In low polar solvents, *Mannich* bases with $\Delta pK_a < 2$ can be characterized as systems with medium to weak hydrogen bonds. $\Delta\nu(OH)$ is in the order of 600–$700\,cm^{-1}$, and the electronic absorption spectra closely resemble those of phenols. No open forms with a free $\nu(OH)$ bands are observed. Vector analysis of dipole moments suggests that the closed forms with intramolecular hydrogen bond prevail in solution. Comparison of the IR and UV-Vis spectra of *Mannich* bases in low-polar solutions with the spectra in the solid state suggests that in such solutions the molecules exist as nonionic forms with intramolecular hydrogen bonds. Dipole moments calculated for such structures precisely fit the dependence of $\Delta\mu$ on ΔpK_a. For systems with $\Delta pK_a > 2$ an increase of $\Delta\mu$ was observed which reaches the value of $2.5\,D$ in benzene at $\Delta pK_a = 4$. At $\Delta pK_a > 3$ an increase of $\Delta\nu(OH)$ was observed, and in the UV-Vis spectra a phenolate absorption appears. Solvent polarity increase and temperature decrease shift the equilibria towards ionic forms. The last fact shows that the enthalpy change upon proton transfer is negative. The observed effect does not depend on concentration. Together with an increase of dipole moments with solvent polarity this indicates that intramolecular proton transfer equilibria exist in solution.

The proton transfer equilibrium constants were determined in different solvents at various temperatures, mainly by UV-Vis spectroscopy, but also with the use of IR spectroscopy. Studies performed simultaneously by various techniques (UV-Vis, IR, NMR spectroscopy, dipole moment measurements) present very uniformly the presence of intramolecular proton transfer in *Mannich* bases. For this process, negative values of ΔH°_{PT} and ΔS°_{PT} were found. Participation of the solvent molecules is evident. It was shown that ΔH°_{PT} correlates linearly with the solvent polarity parameter E_T.

It can be observed that – independent of temperature and solvent polarity – large regions of molecular forms exist. For molecules with ΔpK_a values above 4 the possibility of self-association at room temperature in non-polar and low-polar solvents has been demonstrated. The mechanism of this process is determined by ionization. If the concentration of zwitterionic forms is high, cyclic dimers are formed.

The increase of the polarity of the solvent up to $\varepsilon = 10$ leads to an enhancement of self-association. Polarity increase in nonprotic solvents promotes the proton transfer, but specific interactions with the solute are too weak to break the aggregates. Further aggregation beyond dimers was observed as well. UV-Vis spectra show that all forms in aggregates are ionic species.

Decrease of temperature enhances the tendency to self-association. For example, *Mannich* bases with $\Delta pK_a \approx 3$ which do not associate at room temperature show some evidences of association in CH_2Cl_2 at $200\,K$, whereas compounds with $\Delta pK_a \approx 2.6$ do not associate under such conditions.

Due to the stability of the intramolecular hydrogen bond and the large amount of information obtained by various experimental methods, a general model of hydrogen bonding and proton transfer equilibria in *Mannich* bases was established, pointing out some universal features of hydrogen bonded systems. Informations on the role of the solvent in proton transfer processes or the influence of self-association on the position of proton transfer equilibria in low-polar solvents are obtained and can be applied to more general intermolecular hydrogen bonded systems.

Acknowledgements

The authors want to thank Prof. *P. Schuster* and Prof. *L. Sobczyk* for promoting these studies on *ortho-Mannich* bases, and the *Hochschuljubiläumsstiftung der Stadt Wien* for financial support (Project H 175/95) as well as the ÖAD for financial support within the Polish-Austrian exchange program (17/98). Technical assistance of *A. Wolschann* is also gratefully acknowledged.

References

[1] Part 1: Koll A, Wolschann P (1996) Monatsh Chem **127**: 475
[2] Freedman HH (1961) J Amer Chem Soc **83**: 2900
[3] Sucharda-Sobczyk A, Sobczyk L (1978) Bull Acad Pol Sci Ser Sci Chim **26**: 549
[4] Rospenk M, Fritsch J, Zundel G (1984) J Chem Phys **88**: 321
[5] Rospenk M, Zeegers-Huyskens T (1987) Spectroscopy Letters **20**: 177
[6] Rospenk M, Zeegers-Huyskens T (1986) Spectrochimica Acta **42A**: 499
[7] Rutkowski K, Koll A (1994) J Mol Struct **322**: 195
[8] Jeyaraj M, Sobhanadri J (1980) J Chem Phys **73**: 5766
[9] Jeyaraj M, Sobhanadri J (1980) J Chem Soc Faraday II **76**: 589
[10] Osipov OA, Izmailov CM, Gornovskij AD, Orlova LW, Kaschireninov OE (1935) Zh Obsch Kchim **35**: 265
[11] Rustamova BM, Kolodiazschnyi IW, Osipov OA, Garnovskij AD, Salimov MA, Bilalov SB (1971) J Org Chem USSR **7**: 770
[12] Huyskens P, Zeegers-Huyskens T (1964) J Chim Phys **61**: 81
[13] Sobczyk L, Rospenk M (1980) In: Proceedings of "Analytiktreffen Neubrandenburg". Karl Marx University Leipzig, p 34
[14] Filarowski A, Koll A, Glowiak T (1997) J Chem Cryst **27**: 707
[15] Ratajczak H, Sobczyk L (1969) J Chem Phys **50**: 556
[16] Nouwen R, Huyskens P (1973) J Mol Struct **16**: 459
[17] Hawranek JP, Oszust J, Sobczyk L (1972) J Phys Chem **76**: 2112
[18] Sobczyk L, Engelhardt H, Bunzl K (1976) In: Schuster P, Zundel G, Sandorfy C (eds) The Hydrogen Bond – Recent Developments in Theory and Experiments, vol 3. North Holland, Amsterdam New York Oxford, p 937
[19] Koll A (1983) Bull Soc Chem Belg **92**: 313
[20] Pawelka Z, Rospenk M, Sobczyk L (1987) Bull Soc Chem Belg **96**: 415
[21] Rospenk M, Koll A (1983) Bull Soc Chem Belg **92**: 329
[22] Sobczyk L, Pawelka Z (1973) Roczn Chem **47**: 1523
[23] Rospenk M, Sobczyk L (1989) Magn Res Chem **27**: 445
[24] Martinek H, Wolschann P (1981) Bull Soc Chem Belg **90**: 37
[25] Schreiber VM, Koll A, Sobczyk L (1978) Bull Acad Pol Sci Ser Sci Chim **26**: 651
[26] Köhler G, Wolschann P (1987) J Chem Soc Farad Trans II **83**: 513
[27] Szemik-Hojniak A, Koll A (1993) J Photochem Photobiol A Chem **72**: 123
[28] Köhler G, Wolschann P, Rotkiewicz K (1992) Proc Indian Acad Sci (Chem Sci) **104**: 197
[29] Koll A, Rospenk M, Sobczyk L (1981) J Chem Soc Farad Trans I **77**: 2309
[30] Rospenk M (1990) J Mol Struct **221**: 109
[31] Cumming JB, Keberle P (1978) Canad J Chem **56**: 1
[32] Zundel G, Fritsch J (1984) J Phys Chem **88**: 6295
[33] Rospenk M, Koll A, Sobczyk L (1996) Chem Phys Lett **261**: 283
[34] Rospenk M, Ruminskaya UG, Schreiber VM (1982) Z Prikl Spektrosk **36**: 756
[35] Krämer R, Zundel G, Brzezinski B, Olejnik J (1992) J Chem Soc Faraday Trans **80**: 1659
[36] Brzezinski B, Radziejewski P, Rabolt A, Zundel G (1995) J Mol Struct **355**: 185
[37] Köhler G, Martinek H, Parasuk W, Rechthaler K, Wolschann P (1995) Monatsch Chem **126**: 299

[38] Rospenk M, Zeegers-Huyskens T (1987) J Phys Chem **91**: 3974
[39] Kulbida A, Nosov A, Koll A, Rospenk M, Sobczyk L (1991) J Mol Struct **248**: 217
[40] Rutkowski K, Melikova SM, Koll A (1994) Vibr Spectr **7**: 265
[41] Melikova SM, Rutkowki KS, Shchepkin DN, Koll A (1994) Opt Spectrosk **77**: 684
[42] Brzezinski B, Maciejewska H, Zundel G, Kramer R (1990) J Phys Chem **94**: 528
[43] Brzezinski B, Zundel G, Kramer R (1988) J Mol Struct **198**: 243
[44] Zundel G (1976) In: Schuster P, Zundel G, Sandorfy C (eds) The Hydrogen Bond – Recent Developments in Theory and Experiments, vol 3, chap 15. North Holland, Amsterdam New York Oxford
[45] Perrin DD (1972) Dissociation Constants of Organic Bases in Aqueous Solutions. Butterworth, London.
[46] Filarowski A, Koll A (1996) Vibr Spectr **12**: 15
[47] Filarowski A, Szemik-Hojniak A, Glowiak T, Koll A (1997) J Mol Struct **404**: 67
[48] Filarowski A, Koll A (1998) Vibr Spectr **17**: 123
[49] Wojtowicz A, Malecki J (1980) Bull Acad Pol Sciences **28**: 543
[50] Rospenk M, Koll A (1993) Polish J Chem **67**: 1851
[51] Rospenk M, Koll A, Sobczyk L (1995) J Mol Liquids **67**: 63
[52] Rospenk M (1992) Wiad Chem **46**: 139

Received August 18, 1998. Accepted (revised) February 18, 1999

Competitive Hydrogen Bonds and Conformational Equilibria in 2,6-Disubstituted Phenols Containing two Different Carbonyl Substituents

Alexandra Simperler and **Werner Mikenda**[*]

Institut für Organische Chemie, Universität Wien, A-1090 Vienna, Austria

Summary. Rotational isomers of ten 2,6-disubstituted phenols containing two different carbonyl substituents (COOH, COOCH$_3$, CHO, COCH$_3$, CONH$_2$) have been investigated theoretically with the aid of DFT methods (B3LYP/6-31G(d,p)). The relative stabilities of four to five conformers of each compound were determined by full geometry optimizations for the free molecules and, in order to simulate solvent effects, also for molecules in reaction fields with dielectric constants up to $\varepsilon = 37.5$. Comparison with available experimental IR spectroscopic data revealed excellent agreement with the theoretically predicted stability sequences and conformational equilibria. The two determinative energy contributions, that of the attractive intramolecular hydrogen bond interactions and that of the (mostly repulsive) interactions between the phenolic oxygen atoms and the other non-hydrogen-bonded carbonyl substituents, were separately calculated from isodesmic reactions (benzoyl compound + phenol $\rightleftharpoons$ 2-hydroxybenzoyl compound + benzene). The stability sequences of the conformers, as obtained from the sum of each two energy contributions, almost exactly comply with the sequences obtained from the full geometry optimizations. With all compounds the conformation of the most stable isomer is determined by the energetically most favourable non-bonded O$\cdots$R–C interaction and not by the more favourable one of the two possible O–H$\cdots$O=C H-bond interactions.

Keywords. DFT calculations; Conformational equilibria; 2,6-Disubstituted phenols; Hydrogen bonding; IR spectra; Isodesmic reactions.

Konkurrierende Wasserstoffbrückenbindungen und Konformationsgleichgewichte in 2,6-disubstituierten Phenolen mit zwei unterschiedlichen Carbonylsubstituenten

Zusammenfassung. Die strukturellen, energetischen und spektroskopischen Eigenschaften der Rotationsisomeren von zehn unsymmetrisch substituierten 2,6-Dicarbonylphenolen (COOH, COOCH$_3$, CHO, COCH$_3$, CONH$_2$) wurden mit DFT-Methoden (B3LYP/6-31G(d,p)) untersucht. Die relativen Stabilitäten von vier bis fünf Konformeren jeder Verbindung wurden durch Geometrieoptimierungen einerseits der freien Moleküle und andererseits, um auch Lösungsmitteleffekte zu berücksichtigen, in Reaktionsfeldern mit Dielektrizitätskonstanten bis zu $\varepsilon = 37.5$ bestimmt. Vergleiche mit experimentellen IR-spektroskopischen Daten zeigen eine vortreffliche

---

[*] Corresponding author

Übereinstimmung mit den theoretisch vorhergesagten Stabilitätssequenzen und Konformations-gleichgewichten. Die beiden entscheidenden Energiebeiträge, jene der Wasserstoffbrückenbindungen und jene der (zumeist abstoßenden) Wechselwirkungen zwischen den phenolischen Sauerstoffato-men und den nicht wasserstoffbrückengebundenen Carbonylsubstituenten, wurden mit Hilfe von isodesmischen Reaktionen berechnet (Benzoyl-Verbindung + Phenol $\rightleftharpoons$ 2-Hydroxy-benzoyl-Verbindung + Benzol). Die aus den Summen der beiden Beiträge ermittelten Stabilitätssequenzen stimmen nahezu exakt mit jenen überein, die bei den vollständigen Geometrieoptimierungen erhalten wurden. In allen Fällen wird die Konformation des stabilsten Isomers durch die energetisch günstigste O$\cdots$R–C-Wechselwirkung bestimmt und nicht durch die günstigere der beiden möglichen O–H$\cdots$O=C-Wasserstoffbrückenbindungen.

Introduction

In the course of theoretical and experimental studies on intramolecular hydrogen bonding in small organic model compounds we have investigated a series of 2,6-disubstituted phenols containing two carbonyl substituents. Compounds of this type are found as substructural units of various natural substances like phloroglucinoles [1–4], depsides [5–7], euglobals [8], ochratoxines [9], and of synthetic derivatives with physiological activity [10–11]. Whereas 2,6-disubsti-tuted phenols with two equal carbonyl substituents have been the subject of several investigations [12–17], studies about phenols containing two different carbonyl substituents are very scarce [18, 19]. With respect to intramolecular hydrogen bonding it is, however, just the latter family that offers some particularly challenging questions. Above all, these compounds are capable of forming two (competitive) kinds of intramolecular O–H$\cdots$O=C hydrogen bonds (H-bonds) between the phenolic OH group and the two different carbonyl oxygen atoms. Additionally, for both H-bonded species there exist two rotational isomers with the non-H-bonded carbonyl group being either *syn* or *anti* with respect to the phenolic C–O group. Hence, with any such compound we have to deal with (at least) four different conformers (Fig. 1), and an understanding of the structural and/or spectroscopic properties of a given compound requires knowledge of the factors that govern the stability sequences (*i.e.* the energies or at least the energy differences) of the conformers. Basically, to a first approximation there are two contributions to the total energy of a given conformer that have to be taken into account: (*i*) the attractive H-bond interaction between the phenolic OH group and one of the two carbonyl oxygen atoms and (*ii*) the (mostly repulsive) interaction between the phenolic oxygen atom and the second non-H-bonded carbonyl substituent.

The present study includes ten different 2,6-disubstituted phenols that result from the combinations of each two of five different carbonyl substituents: COOH,

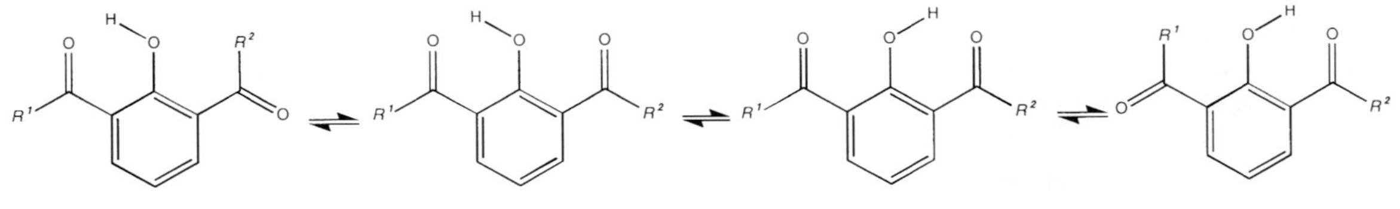

Fig. 1. The four basic conformers of 2,6-disubstituted phenols containing two different carbonyl substituent

$COOCH_3$, CHO, $COCH_3$, and $CONH_2$. From theoretical and experimental structural and spectroscopic data (H-bond energies, bond distances, vibrational frequencies, NMR shifts,...) of 2-hydroxybenzoyl compounds [20, 21] it has been safely established that the H-bond strengths increase within the series $COOH \leq COOCH_3 < CHO < COCH_3 < CONH_2$. This is the starting point of our concerns, since naively one might expect that the H-bond strength should be the main factor in governing the stabilities of the conformers of a given compound. In the following sections, after some experimental and computational details, we firstly focus on theoretical stability sequences as obtained from full geometry optimizations. Data are reported not only for free molecules, but we also account for solvent effects using the *Onsager* reaction field method [22]. Secondly, we confront the theoretical results with experimental IR spectroscopic data in order to assess the reliability of the calculations. Thirdly, we deal with isodesmic reactions, which can be used to evaluate independently the energy contributions of the various kinds of H-bonded and non-bonded interactions. For any given conformer a total interaction energy can be calculated from the sum of two such contributions, and it turns out that the stability sequences obtained in that way almost perfectly agree with the sequences obtained from the full geometry optimizations. Finally, we discuss some pertinent points that emerge from exploring these isodesmic reactions, and we show that this approach can clearly contribute toward the understanding of the conformational stability sequences and, hence, toward the understanding of the intrinsic properties of the title compounds.

Materials and Methods

The compounds included in this study are listed in Table 1 along with the subsequently used compound numbering. IR spectra of CCl_4 and $CDCl_3$ solutions were measured with a Perkin-Elmer Spectrum 2000 FTIR spectrometer. 3-Formyl-2-hydroxy-benzoic acid (**2**) was commercially

Table 1. Compounds and compound numberings

	R^1	R^2
1	-OH	$-OCH_3$
2	-OH	-H
3	-OH	$-CH_3$
4	-OH	$-NH_2$
5	$-OCH_3$	-H
6	$-OCH_3$	$-CH_3$
7	$-OCH_3$	$-NH_2$
8	-H	$-CH_3$
9	-H	$-NH_2$
10	$-CH_3$	$-NH_2$

available from Lancaster. 3-Formyl-2-hydroxy-methylbenzoate (**5**) has been prepared from **2** by standard methods.

The quantum chemical calculations were performed with the Gaussian92 [23] and Gaussian94 [24] programs. Geometric, energetic, and vibrational spectroscopic data of the title compounds and of the components of the isodesmic reactions were computed at the density functional B3LYP level of theory [25, 26] using the 6-31G(d,p) basis set [27]. If not otherwise stated, the optimized geometries were calculated without any constraints. For the simulation of solvent effects, we used the self-consistent *Onsager* reaction field method [22] as implemented in the Gaussian program packages. Within the *Onsager* approach the electrostatic effect of a solvent is represented in terms of the interaction between the molecular dipole moment and a reaction field which is defined by $(2(\varepsilon - 1)/2\varepsilon + 1) \cdot a_0^{-3} \cdot \mu$. Calculations were performed for dielectric constants $\varepsilon = 2.2$, 4.8, and 37.5, corresponding to CCl_4, $CDCl_3$, and CD_3CN solutions, respectively [28]. The radii of the spherical cavities occupied by the molecules, a_0, were calculated from the volumes of the optimized structures of the free molecules and increased by 50 pm in order to account for the *van der Waals* radii of the surrounding solvent molecules.

Results and Discussion

The theoretical energy differences relative to the most stable conformer of a given compound for a total of 44 different conformers are summarized in Table 2. Each first line contains data of free molecules, each second line data of molecules in a

Table 2. Theoretical[a] energy differences ($kJ \cdot mol^{-1}$) between the rotational isomers[b], as obtained from full geometry optimizations (ΔE_{tot}) and from isodesmic reactions (ΔE_{int}, in parentheses) for free molecules (each first line) and for $\varepsilon = 37.5$ (each second line)

	a	**b**	**c**	**d**$_{(E)}$[c]	**d**$_{(Z)}$[c]
1	10(4)	15(7)	14(6)	0(0)	10(3)
	32(22)	34(27)	30(24)	0(0)	27(19)
2	0(0)	15(11)	15(12)	3(6)	10(9)
	1(6)	15(21)	18(24)	0(0)	19(19)
3	3(0)	24(16)	15(7)	0(2)	10(5)
	16(12)	35(32)	26(23)	0(0)	25(19)
4	0(0)	32(26)	28(22)	11(16)	24(19)
	14(0)	45(31)	36(23)	0(0)	35(19)
5	0(0)	15(11)	16(13)	11(10)	
	0(0)	15(15)	22(21)	3(16)	
6	0(0)	22(16)	14(9)	8(6)	
	0(0)	21(20)	15(14)	15(9)	
7	0(0)	32(26)	30(23)	25(20)	
	0(0)	33(31)	29(26)	27(22)	
8	3(4)	25(20)	16(12)	0(0)	
	12(6)	23(26)	10(14)	0(0)	
9	0(0)	33(26)	29(21)	13(9)	
	3(0)	27(31)	13(20)	0(6)	
10	0(0)	34(27)	40(30)	17(14)	
	0(0)	28(31)	28(32)	10(13)	

[a] B3LYP/6-31G(d,p) level of theory; [b] relative to the most stable conformer; [c] (*E*)- and (*Z*)-conformations of non-bonded carboxyl groups (see also Table 5)

reaction field with a dielectric constant of $\varepsilon = 37.5$ corresponding to a solution in CD_3CN. Each first number refers to the relative energy difference between the total energies, ΔE_{tot}, as obtained from full geometry optimizations, each second number (in parentheses) to the relative differences between the interaction energies, ΔE_{int}, as obtained from isodesmic reactions. For the *anti*-conformers of non-bonded carboxyl groups (**1d–4d**) we additionally distinguish between the (*E*)- and the (*Z*)-conformation of the COOH group (**d**$_{(E)}$ and **d**$_{(Z)}$, see also Table 5), the latter being commonly the more stable conformation of carboxylic acids. As should be noted at this point, in order to distinguish between the intrinsic O–H· · O=C H-bond and the non-bonded O· · ·O=C or O· · ·*R*–C interactions, we use the term "non-bonded" also in those few cases where the O· · ·*R*–C interaction is actually also a weak H-bond.

Full geometry optimizations

If we firstly confine to the data obtained from full geometry optimizations, even a quick inspection of Table 2 reveals some general points, three of which should be noted. (1) Referring to the sequence of proton acceptor capabilities noted in the introduction, in only three out of the ten compounds the most stable conformer of the free molecule displays the stronger one of the two possible O–H· · O=C H-bonds (conformers **d** of **1, 3, 8**), whereas in seven cases just the opposite is true (conformers **a** of **2, 4–7, 9, 10**). Obviously, the naive expectation that the strength of the characteristic intramolecular O–H· · O=C H-bond should be mainly determinative for the stability sequence of the rotameric isomers of a given compound does not apply and, consequently, the interaction between the phenolic oxygen and the non-H-bonded carbonyl substituent must play an important or even decisive role in determining that stability sequence. (2) As to the non-bonded interactions, in all instances the most stable conformer is an *anti*-conformer (**a** or **d**) and, moreover, the second stable conformer is also an *anti*-conformer, *i.e.* that with the other O–H· · O=C H-bond. Hence, it can be concluded that any of the O· · ·*R*–C interactions of the *anti*-conformations **a** and **d** is always more favourable than the O· · ·O=C interactions of the corresponding *syn*-conformations **b** or **c**. (3) Concerning the environmental effects, as calculated with the *Onsager* reaction field approach, going from the gas phase to polar solvents results in a more or less distinct stabilization of all conformers. In Table 3, the stabilization energies for $\varepsilon = 37.5$, as defined by the differences between the total energies $E(\varepsilon = 37.5) - E(\varepsilon = 1)$, are given along with the dipole moments of the free species. As should only be noted, for $\varepsilon = 2.2$ (CCl_4 solutions) and $\varepsilon = 4.8$ ($CDCl_3$ solutions) the stabilization energies amount to $39(\pm 2)$ and $69(\pm 2)\%$ respectively, of the values at $\varepsilon = 37.5$ which almost exactly matches the theoretical relations of the *Onsager* model. Moreover, from the data given in Table 3 we obtain an excellent correlation between the stabilization energies and the squares of the dipole moments (linear correlation coefficient: $r = 0.989$) which, because the molecular volumes of the considered compounds are very similar, also complies with the presumptions of the *Onsager* approach. Besides these relations, Table 3 clearly shows that the stabilization energies of the conformers of the single compounds are partly significantly different. Consequently, on going from the gas phase to polar solvents

Table 3. Theoretical[a] stabilization energies (kJ · mol^{-1}) for $\varepsilon = 37.5$[b] (each first line), as obtained from full geometry optimizations and dipole moments (*D*) of free isomers (each second line)

	a	b	c	d$_{(E)}$[c]	d$_{(Z)}$[c]
1	2.7	5.0	7.4	24.0	6.5
	2.6	3.6	4.3	7.5	3.9
2	10.6	12.5	8.8	15.0	3.0
	4.6	5.2	4.4	5.6	2.6
3	7.2	9.2	9.3	20.4	5.1
	4.0	4.8	4.8	6.9	3.5
4	7.4	9.3	13.9	33.0	10.3
	4.0	4.7	5.7	8.3	4.7
5	13.3	13.7	7.8	1.5	
	5.5	5.8	4.5	1.9	
6	9.5	10.1	7.8	3.2	
	4.9	5.2	4.7	2.9	
7	10.1	9.5	11.6	7.5	
	4.9	5.1	5.7	4.2	
8	4.1	12.4	15.7	11.2	
	3.1	5.4	6.1	5.0	
9	3.5	12.8	23.6	19.6	
	2.7	5.4	7.1	6.3	
10	6.7	12.3	18.2	13.6	
	3.9	5.6	6.5	5.5	

[a] B3LYP/6-31G(d,p) level of theory; [b] $E(\varepsilon = 37.5)$–$E(\varepsilon = 1)$; [c] (*E*)- and (*Z*)-conformations of non-bonded carboxyl groups (see also Table 5)

we find more or less distinct changes of the energy differences ΔE_{tot} between the rotational isomers (Table 2) and, hence, corresponding changes of the conformational equilibrium distributions. The most prominent example is compound **4**: whereas **4a** is by far the most stable conformer at $\varepsilon = 1$ (11 kJ · mol^{-1} below **4d**$_{(E)}$), **4d**$_{(E)}$ becomes the most stable conformer at $\varepsilon = 37.5$ (14 kJ · mol^{-1} below **4a**).

Comparison between theoretical and experimental data

In order to check the reliability of the above calculations we briefly confront the theoretical data as obtained from the full geometry optimizations with some available experimental results. In particular, the IR carbonyl stretching frequency region seems to be well suited to distinguish between different conformers. Some relevant theoretical and experimental data are compiled in Table 4. For an assignment of experimentally observed carbonyl stretching frequencies one may either refer to the corresponding DFT calculated frequencies or to experimental frequencies of eligible reference compounds, *i.e.* 2-hydroxy-benzoyl compounds for H-bonded carbonyl groups and benzoyl compounds for non-H-bonded carbonyl groups. From experiment as well as from theory it turns out that the non-bonded carbonyl frequencies of the *syn*-conformers are always distinctly higher than those of the reference benzoyl compounds, whereas the non-bonded carbonyl frequencies

Table 4. Theoretical (first figure)[a] and experimental (second figure)[b] carbonyl stretching frequencies (cm^{-1}) of selected compounds and conformers

	(CO)-CH$_3$	(CO)-H	(CO)-OCH$_3$	(CO)-OH
Benzoyl compound	1776/1692	1796/1709	1799/1729	1818/1743
Salicoyl compound	1713/1646	1733/1668	1736/1683	1755/1696
2a		1782/1699		1755/1699
2d$_{(E)}$		1725/1667		1840/1756
2d$_{(Z)}$		1732/(1667)		1801/(1730)
5a		1781/1698	1738/1681	
5d		1732/	1781/	
6a	1765/1683		1734/1683	
6d	1710/1647		1777/1710	
6c	1710/1647		1816/1735	

[a] B3LYP/6-31G(d,p) level of theory; [b] CCl$_4$ solutions

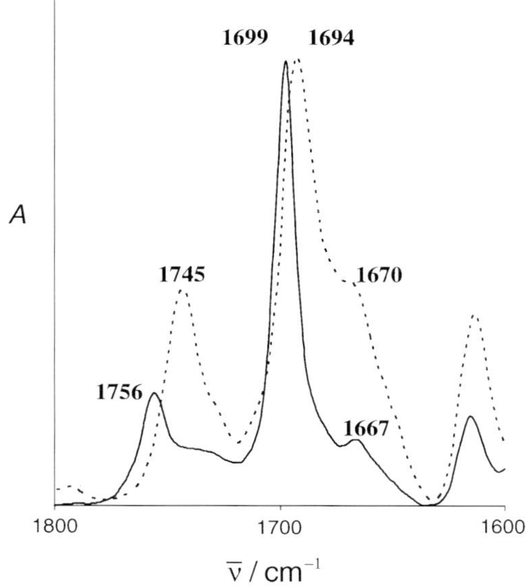

$\overline{\nu}\,/\,cm^{-1}$

Fig. 2. IR spectra of 3-formyl-2-hydroxy-benzoic acid in CCl$_3$ (solid line) and in CDCl$_3$ (dashed line) solutions

of the *anti*-conformers are distinctly lower, with the exception of the anti-(E) conformers of COOH groups (Table 4).

In Fig. 2, the carbonyl stretching frequency region of CCl$_4$ and CDCl$_3$ solution IR spectra of 3-formyl-2-hydroxy-benzoic acid (**2**) are shown. According to the theoretical stability sequence (Table 2) one must expect an equilibrium between two energetically largely similar conformers **2a** and **2d**$_{(E)}$, which excellently complies with the experimental findings: the high frequency band at 1756 cm^{-1} (in CCl$_4$) can reasonably well be assigned to the non-bonded carboxyl group of conformer **2d**$_{(E)}$, the medium frequency band at 1699 cm^{-1} arises from a superposition of the H-bonded carboxyl and the non-bonded formyl groups of conformer **2a**, and the low frequency band at $1667\,\mathrm{cm}^{-1}$ should be due to the H-bonded formyl group of conformer $\mathbf{2d}_{(E)}$. As should be noted, the presence of the two conformers is also confirmed by two corresponding $\nu(\mathrm{OH})$ bands, one at $3525\,\mathrm{cm}^{-1}$ for **2a** and the other at $3395\,\mathrm{cm}^{-1}$ for $\mathbf{2d}_{(E)}$. What is more, according to Table 2 the equilibrium $\mathbf{2a} \rightleftharpoons \mathbf{2d}_{(E)}$ should become shifted towards isomer $\mathbf{2d}_{(E)}$ upon increasing the solvent polarity, and indeed we find that, besides some small frequency shifts, the relative intensities of the carbonyl bands of conformer $\mathbf{2d}_{(E)}$ distinctly increase when going from $\mathrm{CCl_4}$ to $\mathrm{CDCl_3}$ solutions.

For another example, according to Table 2 conformer **5a** should be the most stable rotational isomer of 3-formyl-2-hydroxy-methylbenzoate (**5**), and its stability relative to the next stable conformer $\mathbf{5d}_{(E)}$ increases on increasing solvent polarity. Consistently, we observe a lower frequency band at $1681\,\mathrm{cm}^{-1}$ that can be assigned to the H-bonded carboxylate group and a higher frequency band due to the non-H-bonded formyl group at $1698\,\mathrm{cm}^{-1}$. Moreover, no distinct changes occur on going from $\mathrm{CCl_4}$ to $\mathrm{CDCl_3}$ solution. We note, however, that, although the experimental data agree well with the theoretical prediction, in this special case they do not actually confirm that prediction, since the carbonyl frequencies of the second stable conformer **5d** should be almost the same as those of **5a** (see Table 4).

Finally, we refer to a literature IR spectrum ($\mathrm{CCl_4}$ solution) of 3-acetyl-2-hydroxy-benzoic acid methyl ester (**6**) [19]. In the reported spectrum the carbonyl frequency region is dominated by a strong band at $1683\,\mathrm{cm}^{-1}$ which is obviously due to a superposition of the H-bonded carboxylate group and the non-H-bonded acetyl group of the most stable conformer **6a**. Additionally, the spectrum shows three weak features that can reasonably well be assigned to the two next stable but energetically distinctly less favourable conformers: a band at $1647\,\mathrm{cm}^{-1}$ due to the H-bonded acetyl group of both **6d** and **6c**, a shoulder at $1710\,\mathrm{cm}^{-1}$ due to the non-bonded ester group of the anti-conformer **6d**, and a band at $1735\,\mathrm{cm}^{-1}$ due to the non-H-bonded ester group of the *syn*-conformer **6c**. In summary, for all three compounds the observed spectral features excellently comply with those predicted from the calculated energies and stability sequences of the conformers, which gives strong evidence for the reliability of the calculations.

Interaction energies from isodesmic reactions

As already noted in the introduction, to a first approximation the energy differences between the conformers of a given compound should be governed by two energy contributions: the attractive $\mathrm{O\!-\!H} \cdots \mathrm{O}\!=\!\mathrm{C}$ H-bond interaction and the (mostly repulsive) interaction between the phenolic oxygen atom and the second non-H-bonded carbonyl substituent. For a more detailed understanding of the stability sequences it would be desirable to quantitatively evaluate these energy contributions independently from each other. Here we run into the problem that, other than with intermolecular interactions, the definition and evaluation of intramolecular interaction energies is by far not obvious, because the natural reference systems (i.e. the free molecules) do not exist. With more simple cases, such as *ortho*-substituted phenols, intramolecular H-bond energies are most commonly assessed from the energy differences between the H-bonded species and the conformer with the OH group rotated by $180°$ around the C–O single bond, either

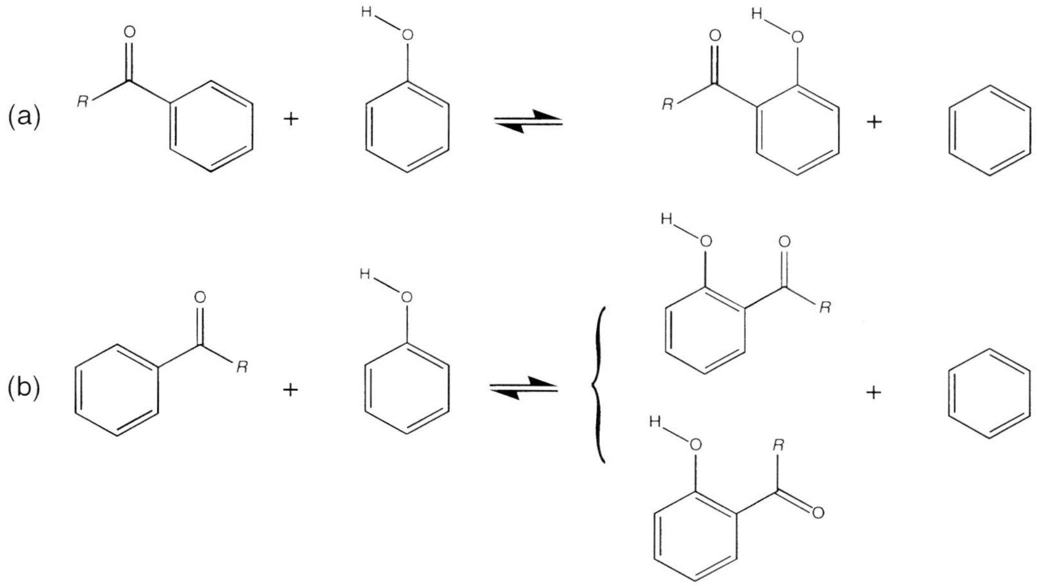

Fig. 3. Isodesmic reactions describing the formation (a) of the intramolecular O–H$\cdots$O=C H-bonds and (b) of the non-bonded O$\cdots$O=C or O$\cdots$R–C interactions

with or without subsequent geometry optimization. For the title compounds, because of the second carbonyl substituent, these possibilities are of course immediately ruled out and, moreover, they are basically not at all applicable for evaluating non-H-bonded interaction energies.

Isodesmic reactions [29] provide an alternate tool for calculating intramolecular interaction energies. The underlying idea is that an intramolecular interaction may formally be built up by a fictitious chemical reaction, and the energy of this interaction-building reaction should be representative of the interaction energy. The isodesmic reactions that should appropriately describe the formation of the intramolecular O–H$\cdots$O=C H-bonds and the formation of the non-bonded O$\cdots$O=C or O$\cdots$R–C interactions are shown in Figs. 3a and 3b, respectively. The theoretical energies that were calculated from the total energies of the four components of the corresponding reaction (E(2-hydroxy-benzoyl) + E(benzol) − E(benzoyl) − E(phenol)) are compiled in Table 5. Within our approach, the sum of the two energy contributions of a given conformer $E_{\mathrm{int}} = E_{\mathrm{Hb}} + E_{\mathrm{nb}}$ (E_{Hb} for the formation of the H-bond and E_{nb} for the formation of the non-bonded interaction) should be representative of the entire interaction energy. The respective energy differences ΔE_{int} (relative to the most stable conformer of a given compound) have already been presented in Table 2 along with the corresponding energy differences as obtained from full geometry optimizations. As can easily be seen, although there are partly some quantitative differences, the stability sequences of the conformers of any given compound as obtained from the two energy scales almost perfectly agree with each other. For the free molecules, in only one out of ten cases (*i.e.* with compound **3**) we find an interchange between the two most stable conformers: E_{tot} prefers **d** by $2\,\mathrm{kJ}\cdot\mathrm{mol}^{-1}$, whereas E_{int} prefers **a** by $3\,\mathrm{kJ}\cdot\mathrm{mol}^{-1}$. Hence, it may

Table 5. Theoretical[a] interaction energies E_{Hb} and E_{nb} (kJ·mol^{-1}) as obtained from isodesmic reactions (see text) for free molecules (each first line) and for $\varepsilon = 37.5$ (each second line) and dipole moments (D) of the free molecules (each third line)

E_{Hb} (O-H··O=C)	E_{nb} (O···R-C)	E_{nb} (O···O=C)
-33	12	18
-32	-6	17
2.1	6.5	2.7
	15	
	13	
	3.3	
-33	16	19
-33	14	19
2.5	3.2	2.3
-38	2	13
-36	-3	11
3.0	4.7	4.2
-39	5	21
-38	0	20
3.2	4.5	3.6
-38	-8	18
-39	-13	18
4.2	4.9	3.8

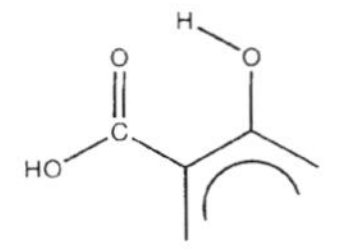

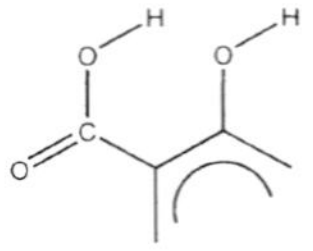

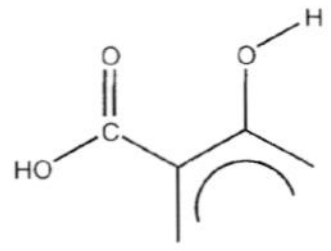

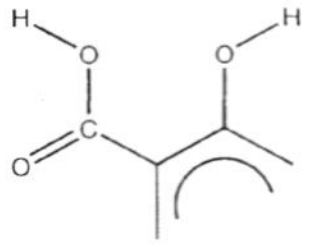

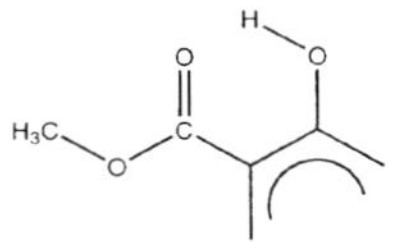

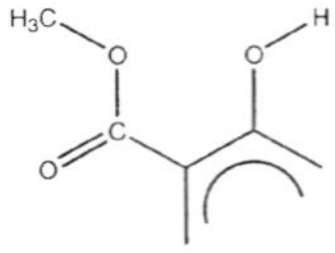

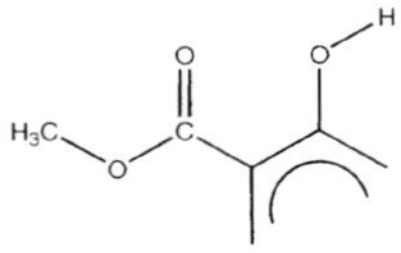

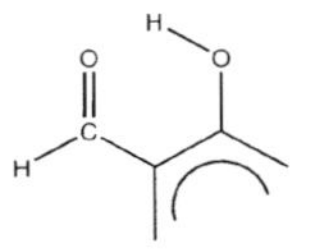

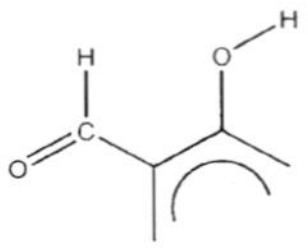

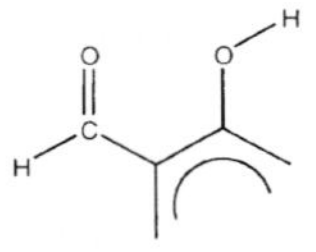

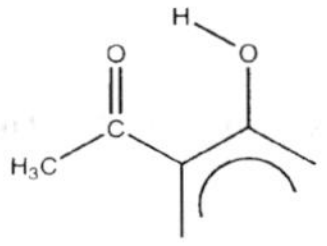

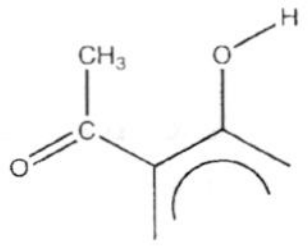

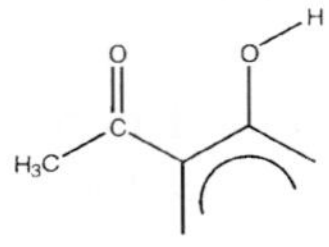

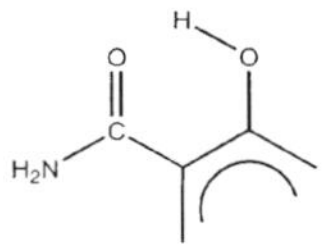

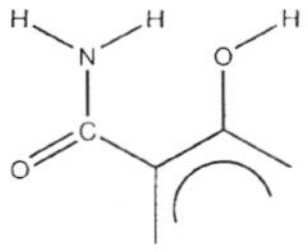

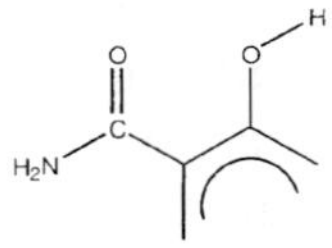

[a] B3LYP/6-31G(d,p) level of theory

certainly be claimed that the stability sequences are well predicted within our isodesmic reaction approach and consequently, this approach should provide a valuable means for a thorough understanding and interpretation of these sequences.

Coming back to the three general points noted above, first of all Table 5 reveals that the energy contributions of the H-bond interactions E_{Hb} are rather similar throughout with a maximum energy difference of only 6 kJ·mol^{-1} between the different carbonyl groups, whereas the contributions of the non-bonded interactions E_{nb} cover a distinctly larger range of about 25 kJ·mol^{-1}. From this finding it is immediately understandable that the conformational stability sequences are primarily governed by the non-bonded and not by the H-bond interactions (*i.e.* the more favourable non-bonded interaction wins over the stronger H-bond interaction). Table 5 indicates that the energies of the non-bonded interactions increase within the series $CONH_2 < COH < COCH_3 < COOH < COOCH_3$, and indeed we find that with almost all compounds the most stable conformer is that with the most favourable non-bonded interaction (the only exception again concerns the two energetically largely similar conformers **a** and **d** of compound **3**).

Secondly, among the non-bonded interactions we can clearly distinguish between the energetically more favourable *anti*-conformations and the less favourable *syn*-conformations, which makes it understandable that the most stable as well as the second most stable conformer of all compounds display an *anti*-conformation. The energy difference between each of the two corresponding *anti*- and *syn*-conformations ranges from 3 to $26 \, kJ \cdot mol^{-1}$ for the free molecules and up to $31 \, kJ \cdot mol^{-1}$ for the most polar environment considered here. Expectedly, the largest difference concerns the amide group, where the *syn*-conformation is associated with a weak but clearly attractive $O \cdots H-N$ H-bond, whereas the smallest difference concerns the ester group where both conformations are associated with a highly unfavourable, strongly repulsive $O \cdots O$ interaction. As a consequence we find that with the most stable conformers of compounds, which contain amide or/and ester groups, the amide group is always engaged in a non-bonded interaction, despite it is the strongest proton acceptor within our series, whereas the ester group almost always acts as proton acceptor, despite it forms rather weak H-bonds.

Thirdly, concerning the solvent effects, as simulated by the *Onsager* reaction field method, on going from the gas phase to CD_3CN solutions Table 5 shows minute or even negligible effects ($\leq 2 \, kJ \cdot mol^{-1}$) for all H-bonded species as well as for all *syn*-conformers and for two *anti*-conformers of the non-H-bonded species. Noticeable effects ($\sim 5 \, kJ \cdot mol^{-1}$) are observed with the *anti*-conformers of amide, acetyl, and formyl groups, whereas the only outstanding case concerns the *anti*-(*E*)-conformation of the carboxyl group, which becomes stabilized by as much as $18 \, kJ \cdot mol^{-1}$. The latter prominent effect agrees well with a common and well known characteristic of carboxylic acids; *e.g.* calculations of benzoic acid at the B3LYP/6-31G(d,p) level of theory reveal that in the gas phase the (*Z*)-isomer is more stable by about $30 \, kJ \cdot mol^{-1}$, whereas the energy difference becomes significantly reduced in polar environments (down to $20 \, kJ/mol$ for $\varepsilon = 37.5$). If we now compare the environmental effects predicted from isodesmic reactions with those obtained from full geometry optimizations (Table 2), we find that the former yield correct trends, whereas the quantitative agreement is less satisfying. The latter drawback is, however, inherently associated with the *Onsager* model: on the one hand, within this approach the electrostatic interaction is directly proportional to the square of the dipole moment of a molecule under consideration, whereas, on the other hand, the dipole moments of the title compounds **1–10** are, of course, not simply sums of the dipole moments of the basic compounds of Table 5.

From the above discussion it may certainly be claimed that the isodesmic reactions are not only remarkably successful in reproducing the conformational stability sequences. In particular, they provide a basis for a quantitative (or at least semiquantitative) understanding of the factors that govern these stability sequences. For the carbonyl groups considered here, a main key is that the sequence of increasing attractive H-bond strengths, as noted in the introduction ($COOH \leq COOCH_3 < CHO < COCH_3 < CONH_2$), and the sequence of decreasing repulsive non-bonded interactions, $COOCH_3 > COOH > COCH_3 > CHO > CONH_2$ (Table 5), are largely similar. Hence, in most cases we are dealing with a rather subtle interplay between two possible H-bond interactions that favour one isomer, and two non-bonded interactions that favour just the other isomer, but the isodesmic reactions can reasonably well help us learn why the stabilities and the stability sequences are as they are.

With respect to further applications, we have recently started to use the isodesmic reaction approach for the prediction of conformational stability sequences of more complex compounds, and preliminary results for some phloroglucin derivatives revealed rather promising results. As examples we notice 3-acetyl-2,4,6-trihydroxy-benzaldehyde, which is a biologically active synthetic analog of grandinol [3], and 2,4-dihydroxy-6-methoxy-3-formylacetophenone. For both compounds, full geometry optimizations and isodesmic reactions yield almost identical results with respect to the conformational stability sequences. In the first case, the energies of the two most stable conformers are largely similar, $\Delta E_{tot} = 2\,kJ \cdot mol^{-1}$ and $\Delta E_{int} = 0\,kJ \cdot mol^{-1}$, whereas in the latter case one conformer is distinctly more stable than the other three conformers: $\Delta E_{tot} > 70\,kJ \cdot mol^{-1}$ and $\Delta E_{int} > 50\,kJ \cdot mol^{-1}$. This gives first strong evidence that, even for more complex compounds, the single energy contributions obtained from isodesmic reactions might actually be used for predictive purposes. Work concerning these issues (the applicability and the limitations) is currently under progress.

Conclusions

A total of 44 rotational isomers of ten 2,6-disubstituted phenols containing two different carbonyl substituents (($C=O$)-R, R=OH, $OOCH_3$, H, CH_3, NH_2) have been investigated theoretically at the B3LYP/6-31G(d,p) level of theory. Calculations were performed for free molecules as well as for reaction fields with $\varepsilon = 2.2$, 4.8, and 37.5. Comparison with available IR spectroscopic data revealed excellent agreement between experiment and theory and confirms the reliability of the calculations. The following are our conclusions:

(a) The energies (and the stability sequence) of the conformers of a given compound are governed by a subtle interplay of an attractive hydrogen bond interaction $O\text{–}H \cdots O{=}C$ between the phenolic OH group and one of the two carbonyl groups, and a (mostly repulsive) interaction $O \cdots R\text{–}C$ or $O \cdots O{=}C$ between the phenolic oxygen atom and the other non-hydrogen-bonded carbonyl substituent.

(b) The energy contributions of the two single interactions can be separately determined from the energies of corresponding interaction-forming isodesmic reactions (benzoyl compound + phenol $\rightleftharpoons$ 2-hydroxybenzoyl compound + benzene). For any conformer, a total interaction energy can be evaluated from the sum of each two contributions, and it appeared that the stability sequences obtained in this way almost perfectly agree with those obtained from full geometry optimizations.

(c) In particular, it turned out that the conformation of the most stable isomer of a given compound is not determined by the more favourable one of the two possible O–H$\cdots$O=C H-bond interactions, but it is always determined by the most favourable non-bonded O$\cdots$$R$–C interaction. The isodesmic reactions have proved remarkably successful in reproducing and predicting the conformational stability sequences. They provide a valuable means for a thorough understanding of the conformational equilibria and of the structural and spectroscopic properties of the compounds.

(d) Finally, evidence is given that the single energy contributions obtained from isodesmic reactions might be also used to predict conformational stabilities of distinctly more complex compounds.

Acknowledgments

The authors thank Prof. *A. Karpfen* for valuable help and discussion. We are grateful to the Computer Centre of the University of Vienna for ample supply of the Digital Alpha 2100 5/375 facilities.

References

[1] Boland DJ, Brophy JJ, Fookes CJR (1992) Phytochemistry **31**: 2178

[2] Miles DH, De Medeiros JMR, Chittawong V, Hedin PA, Swithenbank C, Lidert Z (1991) Phytochemistry **30**: 1131

[3] Bolte ML, Crow WD, Takahashi N, Sakurai A, Uji-Je M, Yoshida S (1985) Agric Biol Chem **49**: 761

[4] Crow WD, Osawa T, Paton DM, Willing RR (1977) Tetrahedron Lett **12**: 1073

[5] Elix JA, Wardlaw JH (1996) Aust J Chem **49**: 539

[6] Elix JA, Chester DO, Gaul KL, Parker JL, Wardlaw JH (1989) Aust J Chem **42**: 1191

[7] Elix JA, Wilkins AL, Wardlaw JH (1987) Aust J Chem **40**: 2023

[8] Takasaki M, Konoshima T, Fujitani K, Yoshida S, Nishimura H, Tokuda H, Nishino H, Iwashima A, Kozuka M (1990) Chem Pharm Bull **38**: 2737

[9] Bredenkamp MW, Dillen JLM, Van Rooyen PH, Steyn PS (1989) J Chem Soc Perkin Trans II, 1835

[10] Yoneyama K, Konnai M, Takematsu T, Iwamura H, Asami T, Takahashi N, Yoshida S (1989) Agric Biol Chem **53**: 1953

[11] Ford RE, Knowles P, Lunt E, Marshall SM, Penrose AJ, Ramsden CA, Summers AJH, Walker JL, Wright DE (1986) J Med Chem **29**: 538

[12] Das R, Mitra S, Nath DN, Mukherjee S (1996) J Chim Phys **93**: 458

[13] Das R, Mitra S, Mukherjee S (1993) Bull Chem Soc Jpn **66**: 2492

[14] Brzezinski B, Zundel G, Krämer R (1987) J Phys Chem **91**: 3077

[15] Golubev NS, Denisov GS (1975) Dokl Akad Nauk SSSR **220**: 1352

[16] Koelle U, Forsen S (1974) Acta Chem Scand A **28**: 531

[17] Tabei M, Tezuka T, Hirota M (1971) Tetrahedron **27**: 301

[18] Reichl G (1996) Thesis, University of Vienna

[19] Hoyer H, Hensel W, Krause G (1965) Z Naturforsch **20**b: 617

[20] Lampert H, Mikenda W, Karpfen A (1996) J Phys Chem **100**: 7418

[21] Steinwender E, Mikenda W (1994) Monatsh Chem **125**: 695

[22] Onsager LJ (1936) J Am Chem Soc **58**: 1486

[23] Frisch MJ, Trucks GW, Schlegel HB, Gill PMW, Johnson BG, Wong MW, Foresman JB, Rob MA, Head-Gordon M, Replogle ES, Gomperts R, Andres JL, Raghavahari K, Binkley JS, Gonzales C, Martin RL, Fox DJ, Defrees DJ, Baker J, Stewart JJP, Pople JA (1993) Gaussian 92, Rev G4; Gaussian Inc, Pittsburgh, PA

[24] Frisch MJ, Trucks GW, Schlegel HB, Gill PMW, Johnson G, Robb MA, Cheeseman JR, Keith T, Petersson GA, Montgomery JA, Raghavachari K, Al-Laham MA, Zakrzewski VG, Ortiz JV, Foresman JB, Peng CY, Ayala PY, Chen W, Wong MW, Andres JL, Replogle ES, Gomperts R, Martin RL, Fox DJ, Binkley JS, Defrees DJ, Baker J, Stewart JP, Head-Gordon M, Gonzales C, Pople JA (1995) Gaussian 94, Rev B3, Gaussian Inc, Pittsburgh, PA

[25] Becke AD (1993) J Chem Phys **98**: 5648

[26] Lee C, Yang W, Parr RG (1988) Phys Rev B **37**: 785

[27] Hehre WJ, Ditchfield R, Pople JA (1972) J Chem Phys **56**: 2257

[28] Maryott AA, Smith EA (1951) Table of Dielectric Constants of Pure Liquids, NBS Circular 14

[29] George P, Bock CW, Trachtman M (1985) J Mol Struct (Theochem) **133**: 11

Received December 10, 1998. Accepted (revised) January 12, 1999

The Dimer of Cyanodiacetylene:
Stacking *vs*. Hydrogen Bonding

Alfred Karpfen

Institut für Theoretische Chemie und Strahlenchemie der Universität Wien, A-1090 Wien, Austria

Summary. The intermolecular energy surface of the cyanodiacetylene dimer was investigated at the MP2 level applying medium to large basis sets. Extensive 2D scans of selected sections of the energy surface were performed. The most stable structure turns out to be an antiparallel stacked dimer. The fully linear structure with a conventional C–H$\cdots$N≡C hydrogen bond is less stable than the antiparallel stacked arrangement by at least $8\,kJ\cdot mol^{-1}$. The intramolecular geometry relaxations relative to the monomer and the vibrational frequency shifts induced by intermolecular interaction are reported. Moreover, the structural and energetic trends in the series $(HCN)_2$, $(HC_3N)_2$, and $(HC_5N)_2$ are discussed.

Keywords. Hydrogen bonding; Cyanodiacetylene; Stacking; MP2 calculations; Theoretical vibrational spectra.

Das Cyanodiacetylendimer: Stackinganordnung oder Wasserstoffbrückenbindung?

Zusammenfassung. Die zwischenmolekulare Energiefläche des Cyanodiacetylendimeren wurde auf MP2-Niveau unter Verwendung von mittleren bis großen Basissätzen untersucht. Ausführliche zweidimensionale punktweise Berechnungen ausgewählter Ausschnitte der Energiefläche wurden durchgeführt. Als stabilste Struktur stellte sich das Dimer mit antiparallel ausgerichteten Monomeren heraus. Die vollständig lineare Struktur mit einer konventionellen C–H$\cdots$N≡C-Wasserstoffbrücke ist um zumindest $8\,kJ\cdot mol^{-1}$ weniger stabil als die antiparallele Anordnung. Die intramolekularen Geometrierelaxationen bezogen auf das Monomere und die durch die zwischenmolekulare Wechselwirkung induzierten Schwingungsfrequenzverschiebungen werden berichtet. Darüber hinaus werden die strukturellen und energetischen Trends in der Reihe $(HCN)_2$, $(HC_3N)_2$ und $(HC_5N)_2$ diskutiert.

Introduction

In the last decade, the gas phase structures and vibrational spectra of hydrogen cyanide and cyanoacetylene clusters have been the subject of intense research, both from the experimental side [1–11] and from theory ([12–15] and references therein). In the case of hydrogen cyanide, the only stable dimer configuration is the fully linear arrangement with a conventional C–H$\cdots$N≡C hydrogen bond [5]. An antiparallel stacked structure with C_{2h} symmetry has recently been found to be a first order saddle point [15]. Similar to the hydrogen cyanide dimer, the experimentally observed cyanoacetylene dimer is also linear [10, 11]. In recent theoretical studies [12, 13], the linear hydrogen-bonded arrangement has indeed

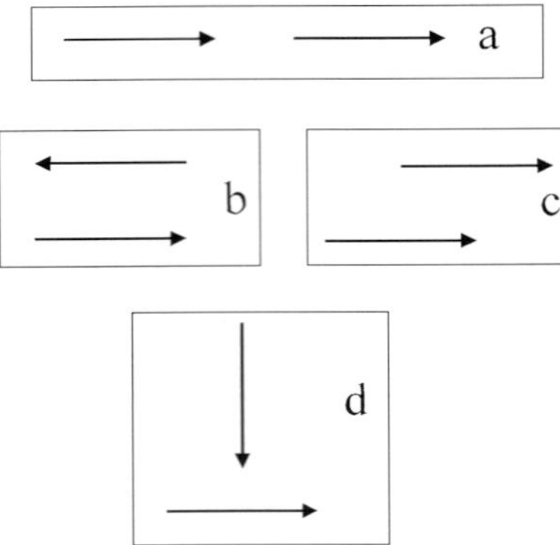

Fig. 1. Schematic drawing of conceivable cyanopolyyne homodimer structures: (a) linear hydrogen-bonded (a), antiparallel stacked (b), slipped parallel stacked (c), and π-type hydrogen-bonded (d)

been found to be the most stable minimum on the cyanoacetylene dimer energy surface. However, in that case an experimentally not yet detected antiparallel stacked structure exhibits also a minimum on the energy surface with a stabilization energy very close to that of the global linear minimum [13].

When comparing the linear and the antiparallel stacked structures sketched in Fig. 1, where the molecules are symbolized simply by arrows, it is immediately evident that the intermolecular interaction energy is dominated mainly by electrostatic contributions in the case of the linear C–H$\cdots$N hydrogen bond, whereas the dispersion energy plays a much more substantial role in the case of the antiparallel stacked structures. From a purely electrostatic point of view, these two orientations are the most stable ones for two point dipoles, with the linear arrangement twice as stable as the stacked configuration. To argue with point dipoles alone is, however, not valid in this case. With increasing chain length of the cyanopolyyne H–(C≡C)$_n$-C≡N the stabilization of the stacked structures, which in addition to the electrostatic contribution to the interaction energy are substantially stabilized by dispersion interactions, is expected to grow much faster than the soon converging stabilization energy of the linear hydrogen-bonded arrangement. To date, there are neither gas phase nor matrix investigations available that deal with the structure of the cyanodiacetylene dimer or the dimers of longer cyanopolyynes. Hence, it appeared to be of interest to test from the theoretical side whether the relative stabilities of the linear and the antiparallel stacked structures are already reversed in the dimer of cyanodiacetylene.

To this end, selected sections of the energy surface of the cyanodiacetylene dimer were scanned with the aid of *ab initio* calculations. In analogy to the previously treated case of cyanoacetylene, the 2D energy surfaces for in-plane translational motions of the antiparallel stacked (Fig. 1b), the parallel stacked (Fig. 1c), and the perpendicular (Fig. 1d) dimers were investigated. The structures and the vibrational spectra of the most stable dimer arrangements were then investigated. The structural changes and the shifts of the vibrational frequencies relative to the cyanodiacetylene monomer are discussed. Particular emphasis is laid on a systematic comparison to analogous data as obtained for the hydrogen cyanide dimer and the cyanoacetylene dimer.

Methods

All quantum chemical calculations were performed with the Gaussian 94 suite of programs [16]. The standard *Møller-Plesset* second order (MP2) [17] method was used in this work since the inclusion of the dispersion interaction at a sufficiently reliable level is vital in this case. While SCF and density functional methods perform quite well for the hydrogen-bonded structures, they are not at all applicable in the case of the stacked structures [13] where an acceptable description of the dispersion interaction is a necessary prerequisite. Guided by the experience gained from previous investigations on hydrogen cyanide [12, 18–21] and cyanoacetylene [12, 13] clusters and to enable a direct comparison, the same basis sets I–IV as used in Ref. [13] were applied also in this work. Basis set I is the 6–31 G(d,p) basis [22, 23]. Basis set II is the $10s6p/6s$ basis set of *Huzinaga* [24, 25] contracted to $6s4p/4s$ and augmented by a set of d functions on nitrogen (0.95) and carbon (1.0) and a set of p functions on hydrogen (0.75). Basis set III is the $11s7p/6s$ *Huzinaga* basis set [24, 25] augmented by two sets of d functions on nitrogen (0.95, 0.3) and carbon (1.0, 0.3) and a set of p functions on hydrogen (0.75). Basis set IV consists of basis set III plus additional flat s, p, and d functions on nitrogen (0.04/0.03/0.1) and carbon (0.03/0.02/0.1) and flat s and 2 sets of p functions on hydrogen (0.03/0.2, 0.05), thus representing overall a contracted $8s6p3d/7s3p$ basis. Basis set IV was used for monomer calculations only.

The 2D scans of the MP2 energy surface of the cyanodiacetylene dimer were performed with basis sets I and II with monomer structures frozen at the respective monomer equilibrium geometries. At the optimized dimer structures, the MP2 interaction energies were corrected for the basis set superposition (BSSE) effect [26] including geometry relaxation. Zero point energy (ZPE) corrections were taken into account at the MP2 level using basis set I only, since the dimer vibrational analysis with the larger basis sets surpassed the available computing resources by far.

Results and Discussion

The monomer

The total energies, the optimized structures, the calculated rotational constants, the dipole moments, and the parallel and perpendicular polarizabilities of the cyanodiacetylene monomer as obtained at the MP2 level using basis sets I–IV are compiled in Table 1. For the purpose of evaluation, analogous data as obtained previously [12, 13] for hydrogen cyanide and cyanoacetylene and a comparison to the best available theoretical and experimental data are included as well. It is evident that all errors present in the description of the monomers will be carried over to the dimers.

The best structure available in the case of the cyanodiacetylene monomer stems from the work of *Botschwina et al.* [27] who performed large basis set coupled cluster (CCSD and CCSD(T)) geometry optimizations with subsequent empirical refinement to reproduce the experimental rotational constant. Internally, the computed MP2 structures as obtained with basis sets I–IV are quite similar. They show, however, some systematic deviations from the structures recommended by *Botschwina et al.* [27]. The computed C≡C and C≡N triple bond distances for all three molecules are too large by about 0.01–0.02 Å, whereas the C–C single bond distances are consistently too short by about 0.005 Å. As with hydrogen cyanide and cyanoacetylene, the computed dipole moment of cyanodiacetylene does not depend too sensitively on the quality of the basis set applied. An important quantity for a reliable treatment of the intermolecular interaction for the case in hand is the

Table 1. MP2 calculated total energies (E), equilibrium structures, rotational constants (B_e), dipole moments (μ), and parallel ($\alpha_\parallel$) and perpendicular ($\alpha_\perp$) polarizabilities of cyanodiacetylene, cyanoacetylene, and hydrogen cyanide

	Basis set				Previous theoretical
	I	II	III	IV	and experimental data[a]
Cyanodiacetylene					
E (hartree)	−245.01404	−245.11301	−245.18362	−245.18610	−
$r(H–C_1)$(Å)	1.0641	1.0629	1.0653	1.0654	1.0626
$r(C_1{\equiv}C_2)$ (Å)	1.2270	1.2160	1.2205	1.2208	1.2091
$r(C_2–C_3)$ (Å)	1.3659	1.3588	1.3608	1.3608	1.3661
$r(C_3{\equiv}C_4)$ (Å)	1.2334	1.2217	1.2266	1.2268	1.2127
$r(C_4–C_5)$ (Å)	1.3659	1.3626	1.3650	1.3653	1.3710
$r(C_5{\equiv}N)$ (Å)	1.1901	1.1778	1.1818	1.1818	1.1613
B_e (GHz)	1.3063	1.3257	1.3184	1.3181	1.3301
μ (D)	4.337	4.329	4.428	4.417	4.407
$\alpha_\parallel$ (Å^3)	19.026	19.818	20.839	21.001	−
$\alpha_\perp$ (Å^3)	2.776	3.508	4.664	5.473	−
Cyanoacetylene[b]					
E (hartree)	−169.08633	−169.15923	−169.20460	−169.20614	−
$r(H–C_1)$(Å)	1.0640	1.0629	1.0652	1.0652	1.0624
$r(C_1{\equiv}C_2)$ (Å)	1.2236	1.2116	1.2159	1.2162	1.2058
$r(C_2–C_3)$ (Å)	1.3770	1.3701	1.3728	1.3729	1.3764
$r(C_3{\equiv}N)$ (Å)	1.1875	1.1752	1.1789	1.1789	1.1605
B_e (GHz)	4.454	4.524	4.499	4.498	4.549
μ (D)	3.712	3.723	3.778	3.769	3.724[c]
$\alpha_\parallel$ (Å^3)	8.902	9.484	9.901	9.975	9.715[d]
$\alpha_\perp$ (Å^3)	1.980	2.469	3.258	3.816	3.722[d]
Hydrogen cyanide[e]					
E (hartree)	−93.16617	−93.21234	−93.23288	−93.23364	−
$r(H–C)$(Å)	1.0653	1.0647	1.0666	1.0666	1.0650
$r(C{\equiv}N)$ (Å)	1.1778	1.1655	1.1686	1.1686	1.1532
B_e (GHz)	42.949	43.730	43.509	43.509	44.278
μ (D)	2.898	2.990	3.019	3.013	3.012
$\alpha_\parallel$ (Å^3)	2.869	3.174	3.262	3.308	3.283[d]
$\alpha_\perp$ (Å^3)	1.133	1.377	1.793	2.084	2.034[d]

[a] Theoretical structural data from Refs. [27–29]; [b] data from Refs. [12, 13]; [c] Ref. [30]; [d] MBPT2 data from Ref. [31]; [e] data partially from Ref. [12]

polarizability and in particular, the anisotropy of the polarizability. The variation of the computed parallel component of the polarizability, *i.e.* the polarizability along the long molecular axis, is modest only. For cyanodiacetylene and cyanoacetylene, using the smallest basis set, a value for $\alpha_\parallel$ is obtained which is only about 10 percent lower than the basis set IV result. In case of hydrogen cyanide, the corresponding deviation amounts to about 15 percent. Of decisive importance for a correct description of the energetics of the stacked structures is, however, the

Table 2. MP2 calculated harmonic vibrational frequencies (cm^{-1}) and infrared intensities $(\text{km} \cdot \text{mol}^{-1})$ of the cyanodiacetylene monomer.

Basis set	I	III	IV	CCSD (T)[a]	CCSD (T)[a,b]
Stretching modes					
ω_1	3527 (104)[c]	3456 (106)	3453 (104)	3455 (91)	3349 (49)
ω_2	2250 (50)	2209 (49)	2208 (48)	2318 (66)	2319 (39)
ω_3	2164 (4.1)	2132 (1.4)	2130 (1.1)	2225 (3.9)	2198 (5.5)
ω_4	2039 (0.2)	2013 (0)	2013 (0.01)	2090 (0.03)	2067 (0.08)
ω_5	1185 (0.02)	1164 (0.01)	1159 (0.02)	1157 (0.13)	1147 (0.1)
ω_6	618 (0.6)	608 (0.5)	606 (0.5)	606 (0.34)	602 (0.34)
Bending modes					
ω_7	590 (70)	595 (89)	566 (77)	–	–
ω_8	503 (12)	409 (7.3)	407 (8.5)	–	–
ω_9	460 (5.3)	312 (3.0)	274 (1.3)	–	–
ω_{10}	263 (11)	209 (7.4)	207 (8.2)	–	–
ω_{11}	115 (1.4)	99 (0.6)	94 (0.7)	–	–

[a] Ref. [27]; [b] fundamental frequencies; [c] infrared intensities in parentheses

accurate value of the perpendicular component of the polarizability, $\alpha_\perp$. Comparing basis sets I and IV, it is obvious that for all three molecules only about 50 percent of $\alpha_\perp$ are obtained when using basis set I. MP2 structure optimizations of the cyanodiacetylene dimer with basis set IV are unfortunately prohibitively time consuming. However, the trends in the computed polarizability data shown in Table 1 should allow a reasonable estimate of the errors still present when comparing the relative energetics of the different minima.

Computed harmonic vibrational frequencies of the cyanodiacetylene monomer are collected in Table 2 and compared to the CCSD(T) results on the stretching modes reported by *Botschwina et al.* [27]. In line with the already observed too large triple bond distances, the MP2 frequencies for the triple bond stretches (ω_2, ω_3, and ω_4) are all too low.

The dimer

1) 2D scans

For a first exploration of the conceivable non-hydrogen bonded configurations, three different 2D sections of the energy surface of the cyanodiacetylene dimer have been scanned. These correspond to the molecular orientations shown schematically in Figs. 1b–d. Contour plots as computed from the MP2/II energies obtained on a regular mesh with Δz increments of 0.5 Å and Δx increments of 0.25 Å are shown in Figs. 2–4. Non-planar orientations were not considered and no BSSE corrections were computed.

By analogy to the already known case of cyanoacetylene [13], the most stable non-hydrogen bonded configuration should be the antiparallel stacked arrangement

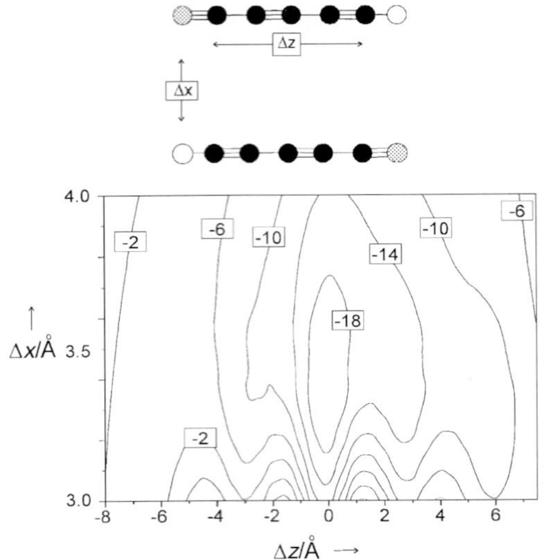

Fig. 2. Contour plot for x,z translations of a cyanodiacetylene monomer in antiparallel orientation to a fixed cyanodiacetylene molecule; energy values obtained from MP2 calculations applying basis set II; contour labels in $kJ \cdot mol^{-1}$ relative to twice the monomer energy; the upper picture indicates $\Delta z = 0$

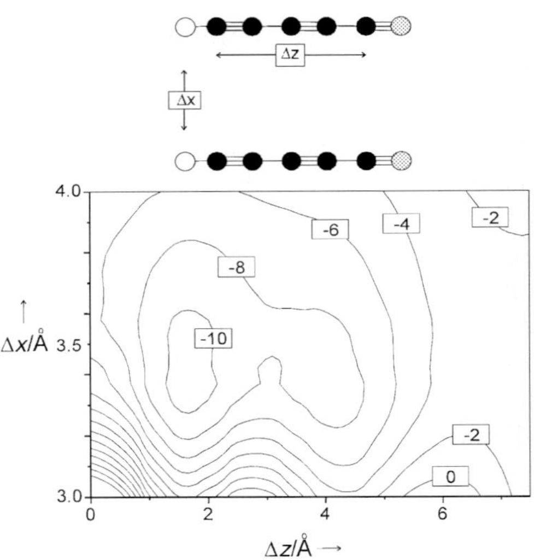

Fig. 3. Contour plot for x,z translations of a cyanodiacetylene monomer in parallel orientation to a fixed cyanodiacetylene molecule; energy values obtained from MP2 calculations applying basis set II; contour labels in $kJ \cdot mol^{-1}$ relative to twice the monomer energy; the upper picture indicates $\Delta z = 0$

in which the two molecular dipoles are oriented such that the hydrogen and nitrogen atoms of different monomers have the closest contact. This results in two strongly distorted, but very weak hydrogen bonds with two nearly 90° bond angles at H and N in the C–H$\cdots$N≡C moiety. It is indeed a matter of taste whether this strongly non-linear C–H$\cdots$N≡C arrangement is still considered as a hydrogen

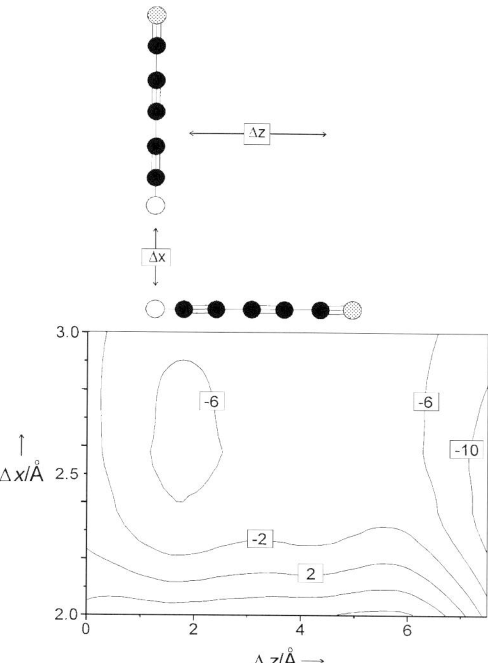

Fig. 4. Contour plot for x,z translations of a cyanodiacetylene monomer in perpendicular orientation to a fixed cyanodiacetylene molecule; energy values obtained from MP2 calculations applying basis set II; contour labels in $kJ \cdot mol^{-1}$ relative to twice the monomer energy; the upper picture indicates
$$\Delta z = 0$$

bond. Even with frozen monomers the computed contour plot depicted in Fig. 2 shows a deep minimum of about $-20\,kJ \cdot mol^{-1}$ for a non-slipped antiparallel stacked structure with an intermolecular distance of about 3.4 Å, qualitatively quite similar to the case of cyanoacetylene, but, when viewed quantitatively, distinctly more attractive. This stabilization is already comparable to that of the linear hydrogen-bonded structure for which by analogy to the cyanoacetylene and hydrogen cyanide dimers a stabilization energy close to $-20\,kJ \cdot mol^{-1}$ could be expected. There are no distinct side minima corresponding to slipped antiparallel stacked conformations.

The contour plot for the parallel stacked case shown in Fig. 3 exhibits two minima, one for a structure in which one monomer is displaced by about 2 Å relative to the other and a second, less stable and shallower minimum for a displacement of about 4 Å. The bare, uncorrected interaction energies for these two structures are in the vicinity of $-10\,kJ \cdot mol^{-1}$.

The contour plot for the perpendicular configuration shown in Fig. 4 displays an extremely shallow minimum corresponding to a π-type hydrogen bond to the $C_1 \equiv C_2$ triple bond with an interaction energy of about $-6\,kJ \cdot mol^{-1}$. There are no corresponding minima for π-type hydrogen bonds to the $C_3 \equiv C_4$ and $C_5 \equiv N$ triple bonds, since the nitrogen atom creates a too attractive basin. Even the configuration where the hydrogen bond donating monomer is rotated by 90° out of its optimal linear structure is more attractive than any of the π-type hydrogen-bonded

80
A. Karpfen

alternatives. Thus, the two latter configurations, slipped parallel stacked and π-type hydrogen-bonded, cannot compete in stability with the non-slipped antiparallel stacked and the conventional linear hydrogen-bonded structures. This situation is entirely analogous to the cyanoacetylene dimer energy surface [13].

Table 3. MP2/I and MP2/III optimized geometrical parameters of the linear hydrogen-bonded $C_{\infty v}$ and the antiparallel C_{2h} stacked structures of the cyanodiacetylene, cyanoacetylene, and hydrogen cyanide homodimers[a]

		$C_{\infty v}$			C_{2h}		
		I	III		I	III
$(HC_5N)_2$	$H–C_1$	1.0644	1.0656	$H–C_1$	1.0650	1.0665
	$C_1{\equiv}C_2$	1.2271	1.2205	$C_1{\equiv}C_2$	1.2283	1.2224
	$C_2–C_3$	1.3648	1.3598	$C_2–C_3$	1.3638	1.3580
	$C_3{\equiv}C_4$	1.2337	1.2269	$C_3{\equiv}C_4$	1.2348	1.2288
	$C_4–C_5$	1.3674	1.3631	$C_4–C_5$	1.3675	1.3626
	$C_5{\equiv}N$	1.1888	1.1808	$C_5{\equiv}N$	1.1908	1.1831
	$N{\cdots}H'$	2.2478	2.2525	$N{\cdots}H'$	3.1862	3.1357
	$H'–C_1'$	1.0697	1.0712	$C_1{\cdots}C_5'$	3.3289	3.2592
	$C_1'{\equiv}C_2'$	1.2285	1.2222	$C_2{\cdots}C_4'$	3.4115	3.3463
	$C_2'–C_3'$	1.3645	1.3597	$C_3{\cdots}C_3'$	3.4456	3.3747
	$C_3'{\equiv}C_4'$	1.2341	1.2274	$\angle HC_1C_2$	177.6	179.1
	$C_4'–C_5'$	1.3685	1.3643	$\angle C_1C_2C_3$	179.5	178.6
	$C_5'{\equiv}N'$	1.1905	1.1821	$\angle C_2C_3C_4$	178.3	178.6
				$\angle C_3C_4C_5$	178.8	179.1
				$\angle C_4C_5N$	177.9	178.2
				$\angle C_5NH'$	91.9	91.5
				$\angle NH'C_1'$	95.4	94.9
$(HC_3N)_2^b$	$H–C_1$	1.0645	1.0656	$H–C_1$	1.0649	1.0661
	$C_1{\equiv}C_2$	1.2235	1.2159	$C_1{\equiv}C_2$	1.2243	1.2170
	$C_2–C_3$	1.3754	1.3712	$C_2–C_3$	1.3757	1.3712
	$C_3{\equiv}N$	1.1869	1.1776	$C_3{\equiv}N$	1.1878	1.1794
	$N{\cdots}H'$	2.2433	2.2532	$N{\cdots}H'$	3.2800	3.2222
	$H'–C_1'$	1.0694	1.0710	$C_1{\cdots}C_3'$	3.3661	3.3161
	$C_1'{\equiv}C_2'$	1.2249	1.2175	$C_2{\cdots}C_2'$	3.4041	3.3747
	$C_2'–C_3'$	1.3762	1.3720	$\angle HC_1C_2$	178.6	179.3
	$C_3'{\equiv}N'$	1.1878	1.1792	$\angle C_1C_2C_3$	178.7	177.7
				$\angle C_2C_3N$	177.8	177.2
				$\angle C_5NH'$	92.0	91.8
				$\angle NH'C_1'$	92.2	93.0
$(HCN)_2^c$	$H–C$	1.0655	1.0673	$H–C$	1.0655	1.0674
	$C{\equiv}N$	1.1758	1.1670	$C{\equiv}N$	1.1776	1.1685
	$N{\cdots}H'$	2.2203	2.2342	$N{\cdots}H'$	3.2842	3.2711
	$H'–C'$	1.0706	1.0729	$C{\cdots}C'$	3.3216	3.3013
	$C'{\equiv}N'$	1.1779	1.1692	$\angle HCN$	178.6	178.7
				$\angle CNH'$	95.1	93.8
				$\angle NH'C'$	86.3	87.6

[a] Bond distances in Å, bond angles in degrees; [b] taken from Refs. [12, 13]; [c] data for the linear structure taken from Ref. [12]

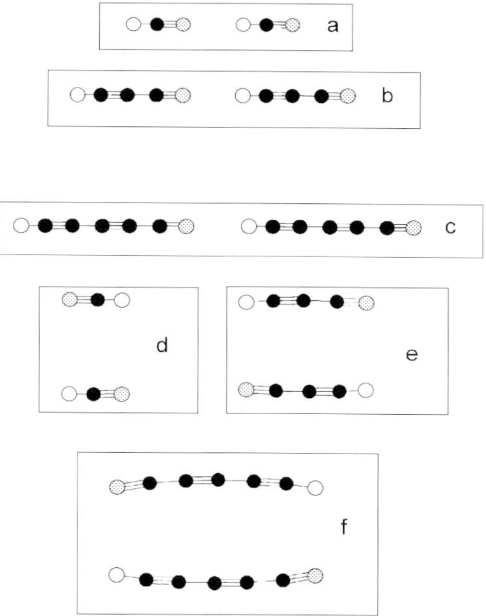

Fig. 5. Schematic drawing of linear hydrogen-bonded (a–c) and antiparallel stacked (d–f) structures of hydrogen cyanide, cyanoacetylene, and cyanodiacetylene dimers

2) Optimized dimer structures

As a consequence of the relative stabilities just discussed, only the two most stable configurations are considered for more detailed structural and vibrational analysis. The optimized geometrical parameters of the linear hydrogen-bonded configuration and of the antiparallel stacked structure of the cyanodiacetylene dimer as obtained with basis sets I and III are compiled in Table 3, where a comparison is also made to the corresponding cyanoacetylene dimer and hydrogen cyanide dimer structures. The structures of these dimers are sketched in Fig. 5. With the exception of the antiparallel stacked hydrogen cyanide dimer, which is a first order saddle point, all other structures are true minima. Comparing the linear hydrogen-bonded structures of the three dimers, it can be observed that, with increasing size of the cyanopolyyne, the local hydrogen bond geometry does not change very much and appears to be very soon converged. A slight, almost insignificant widening of the $H \cdots N$ distance by almost $0.02\,\text{Å}$ occurs upon going from the hydrogen cyanide dimer to the cyanoacetylene dimer. The $H \cdots N$ distance in the cyanodiacetylene dimer is already practically identical to that in the cyanoacetylene dimer.

In the equilibrium structure of the stacked antiparallel cyanodiacetylene dimer the interacting monomers deviate significantly from linearity. Whereas the MP2/III computed $H \cdots N'$ distance amounts to $3.13\,\text{Å}$, the distance between the two central carbon atoms $C_3 \cdots C_3'$ is $3.37\,\text{Å}$, very close to the expected *van der Waals* distance of about $3.4\,\text{Å}$. Again, this structure is very similar to its counterpart in the antiparallel stacked cyanoacetylene dimer and even to the saddle point structure in the stacked hydrogen cyanide dimer.

Table 4. Computed MP2 stabilization energies (ΔE), BSSE-corrected stabilization energies (ΔE(BSSE)), and ZPE-corrected stabilization energies (ΔE(ZPE)), for the linear $C_{\infty v}$ and antiparallel stacked C_{2h} structures of the cyanodiacetylene, cyanoacetylene, and hydrogen cyanide dimers; all values in $kJ \cdot mol^{-1}$

			Basis set		
			I	II	III
$(HC_5N)_2$	$C_{\infty v}$	ΔE	-20.3	-16.5	-18.0
		ΔE(BSSE)	-16.6	-14.9	-17.2
		ΔE(ZPE)	-17.1	-13.3[a]	-14.8[a]
		ΔE(BSSE+ZPE)	-13.4	-11.8	-14.0
	C_{2h}	ΔE	-21.9	-23.5	-25.7
		ΔE(BSSE)	-11.9	-12.3	-22.9
		ΔE(ZPE)	-21.0	-22.6[a]	-24.8[a]
		ΔE(BSSE+ZPE)	-11.0	-11.5	-22.0
$(HC_3N)_2$[b]	$C_{\infty v}$	ΔE	-20.6	-18.0	-18.2
		ΔE(BSSE)	-16.9	-16.5	-17.5
		ΔE(ZPE)	-17.2	-15.9	-15.9
		ΔE(BSSE+ZPE)	-13.6	-14.4	-15.2
	C_{2h}	ΔE	-15.3	-15.9	-16.8
		ΔE(BSSE)	-8.7	-9.3	-15.2
		ΔE(ZPE)	-14.0	-14.8	-15.5
		ΔE(BSSE+ZPE)	-7.4	-8.2	-13.9
$(HCN)_2$[c]	$C_{\infty v}$	ΔE	-22.1	-19.4	-19.3
		ΔE(BSSE)	-18.0	-18.0	-18.7
		ΔE(ZPE)	-17.5	-16.2	-16.1
		ΔE(BSSE+ZPE)	-13.4	-14.8	-15.5
	C_{2h}	ΔE	-11.3	-11.0	-11.6
		ΔE(BSSE)	-7.7	-8.5	-11.0
		ΔE(ZPE)	-9.8	-10.1	-10.7
		ΔE(BSSE+ZPE)	-6.3	-7.6	-10.1

[a] ZPE corrections taken over from MP2/I; [b] data taken from Ref. [13]; [c] data taken from Ref. [12]

3) Dimer stabilization energies

The raw dimer stabilization energies and the corresponding BSSE and ZPE corrected values are collected in Table 4. Turning first to the linear $C_{\infty v}$ structures, we observe a slight increase of the stabilization energy upon increasing the length of the cyanopolyyne. The MP2/III results lead to uncorrected stabilization energies ranging from -19.3 to $-18.0\,kJ \cdot mol^{-1}$ and to BSSE and ZPE corrected values from -14.0 to $-15.5\,kJ \cdot mol^{-1}$. With basis set III the BSSE error amounts only to about 0.6–$0.8\,kJ \cdot mol^{-1}$.

As expected, in the case of the antiparallel stacked C_{2h} structures the stabilization energy is substantially lowered upon going from the hydrogen cyanide dimer *via* the cyanoacetylene dimer to the cyanodiacetylene dimer. The BSSE and ZPE corrected stabilization energies as obtained at the MP2/III level are -10.1, -13.9, and $-22.0\,kJ \cdot mol^{-1}$, respectively. Because of the still large BSSE error when using basis sets I and II and because of the still significantly underestimated perpendicular component of the monomer polarizability when using these two smaller basis sets, the naive routine application of the BSSE correction is not advisable in this case. The result would be a wrong energetic order of the two alternative structures.

Overall, the BSSE corrections tend to favor the linear structures, a consequence of the smaller molecular overlap in the case of the linear structures. The ZPE corrections, on the other hand, tend to favor the stacked structures, a consequence of the larger intermolecular distance and hence smaller perturbations of the intermolecular structure and vibrational frequencies in the stacked structures. Whereas in case of the cyanoacetylene dimer the linear $C_{\infty v}$ structure is still slightly more stable than the C_{2h} configuration, the relative stabilities are clearly reversed in the case of cyanodiacetylene. With about $-22.0\,\mathrm{kJ\cdot mol^{-1}}$, the antiparallel stacked structure is about $8\,\mathrm{kJ\cdot mol^{-1}}$ lower than the stabilization of the linear cyanodiacetylene dimer. Based on the still sizeable error in the perpendicular monomer polarizability at the MP2/III level, it is to be expected that with still more extended basis sets the antiparallel stacked structure is energetically even more preferred.

4) Dimer vibrational frequencies

The MP2/I computed vibrational frequencies of the linear cyanodiacetylene dimer are compiled in Table 5, that of the antiparallel stacked structure are reported in Table 6. More important than the absolute frequency values are the frequency shifts relative to the monomer vibrations. The latter are expected to be quite reliable within a few wavenumbers.

In the case of the hydrogen cyanide dimer and of larger hydrogen cyanide clusters, mainly the spectral region of the C–H stretching frequencies has been investigated from the experimental side [1–6, 8, 9], with the exception of CARS and PARS studies in the region of the $C\equiv N$ stretching frequency [1, 2]. For the cyanoacetylene clusters only the C–H stretching region has been scanned so far by infrared spectroscopic experiments [10–11]. However, for both cases the two dimer C–H stretching frequencies, one corresponding to the non-hydrogen-bonded C–H group, the other to its hydrogen-bonded counterpart, have been accurately determined. As already mentioned, to date no experimental data are available for the case of the cyanodiacetylene dimer.

For the C–H stretching frequencies of the linear cyanodiacetylene dimer we compute shifts of $-3\,\mathrm{cm^{-1}}$ for the non-hydrogen-bonded C–H group and $-73\,\mathrm{cm^{-1}}$ for the hydrogen-bonded C–H group. At the same methical level the corresponding shifts in the linear hydrogen cyanide and cyanoacetylene dimers [12] are -1 and $-74\,\mathrm{cm^{-1}}$, and -4 and $-72\,\mathrm{cm^{-1}}$, respectively, which may be compared to the corresponding experimental shifts of -4 and $-70\,\mathrm{cm^{-1}}$ in the case of the hydrogen cyanide dimer, and -3 and $-66\,\mathrm{cm^{-1}}$ for the cyanoacetylene dimer. Thus, the C–H stretching frequency shifts for the linear cyanodiacetylene dimer should be almost identical to that observed in the linear cyanoacetylene dimer, provided that this configuration can be observed experimentally. Compared to the infrared intensity of the monomer C–H stretch, the intensity of the lower-

Table 5. Computed MP2/I harmonic vibrational frequencies (cm^{-1}) and infrared intensities $(km \cdot mol^{-1})$ of the linear $C_{\infty v}$ cyanodiacetylene dimer

Frequencies	Infrared intensities	Frequencies	Infrared intensities
Intramolecular stretchings		Intramolecular bendings	
3524 (−3)[a]	110	757 (167)	75
3454 (−73)	715	594 (4)	77
2250 (0)	62	523 (20)	6
2247 (−3)	176	499 (−4)	19
2168 (4)	3	482 (22)	2
2160 (−4)	37	444 (−16)	1
2041 (2)	2	282 (19)	7
2033 (−6)	19	248 (−15)	10
1191 (6)	0.4	130 (15)	1
1187 (2)	2	115 (0)	3
627 (9)	0.2		
619 (1)	0.3		
Intermolecular stretching		Intermolecular bendings	
71	2	43	4
		8	0.2

[a] Frequency shifts relative to the cyanodiacetylene monomer in parentheses

lying dimer C–H stretch is increased by about a factor of 7, an increase typical for medium-strength hydrogen bonds, whereas that of the higher-lying C–H stretch has essentially the same intensity as that of the monomer since it originates from the non-hydrogen-bonded C–H group.

For the nearly degenerate C–H stretching frequencies of the antiparallel stacked cyanodiacetylene dimer we predict shifts of $-9\,cm^{-1}$. For symmetry reasons, only the b_u mode is infrared-active with an intensity of about twice that of the monomer. This shift of $-9\,cm^{-1}$ is comparable and only slightly larger than the corresponding shift of $-7\,cm^{-1}$ for the hitherto unobserved antiparallel stacked cyanoacetylene dimer, if computed at the same level of approximation [13].

Inspecting the remaining dimer frequency shifts, we observe that the H–C≡C bending modes are also interesting candidates to eventually discern between the two alternatives. For the linear cyanodiacetylene dimer shifts of 167 and $4\,cm^{-1}$ are computed. The larger shift originates again from the hydrogen-bonded C≡C–H group. For the antiparallel stacked dimer the computed shifts of $+19$ and $-12\,cm^{-1}$ lead to a quite different pattern. Although there are sizeable shifts in other spectral regions as well, they are probably less suited to allow for a reliable discrimination between the two structural alternatives.

Table 6. Computer MP2/I harmonic vibrational frequencies (cm^{-1}) and infrared intensities ($km \cdot mol^{-1}$) of the antiparallel stacked C_{2h} cyanodiacetylene dimer

Frequencies	Infrared intensities	Frequencies	Infrared intensities	Frequencies	Infrared intensities
Intramolecular stretchings		In-plane intramolecular bendings		Out-of-plane intramolecular bendings	
$3518\,(-9)^{a}$	0	$578\,(-12)$	106	$609\,(19)$	0
$3518\,(-9)$	202	$578\,(-12)$	0	$609\,(19)$	76
$2245\,(-5)$	113	$499\,(-4)$	0	$511\,(8)$	0
$2238\,(-12)$	0	$497\,(-6)$	24	$511\,(8)$	14
$2160\,(-4)$	0	$428\,(-32)$	7	$477\,(17)$	0
$2160\,(-4)$	10	$418\,(-42)$	0	$477\,(17)$	0.4
$2033\,(-6)$	0	$244\,(-19)$	0	$268\,(5)$	10
$2033\,(-6)$	2	$242\,(-21)$	14	$266\,(3)$	0
$1189\,(4)$	0.2	$125\,(10)$	1	$121\,(6)$	0
$1188\,(3)$	0	$121\,(6)$	0	$116\,(1)$	1
$620\,(2)$	0				
$620\,(2)$	0.4				
Intermolecular stretching		In-plane intermolecular bendings		Out-of-plane intermolecular bendings	
82	0	69	9	25	4
		43	0		

[a] Frequency shifts relative to the cyanodiacetylene monomer in parentheses

Conclusions

A large-scale systematic study of the energy surface of the cyanodiacetylene dimer has been performed with a subsequent structural and vibrational spectroscopic characterization of the two energetically most stable minima. Contrary to the cases of hydrogen cyanide and cyanoacetylene dimers, the most stable cyanodiacetylene dimer has an antiparallel stacked structure with an interaction energy of about -20 to $-24\,kJ\,mol^{-1}$. The linear hydrogen-bonded structure is less stable by at least $8\,kJ \cdot mol^{-1}$. The attractive contribution of the dispersion energy which increases with increasing length of the cyanopolyyne chain leads to a reversal of the relative stabilities of linear and stacked structures upon going from cyanoacetylene to cyanodiacetylene. From the trends just discussed in the series of short cyanopolyynes it is to be expected that in larger cyanopolyynes the antiparallel stacked structures will always be more stable and that it will be quite difficult if not impossible to observe the conventional linear hydrogen-bonded configurations experimentally. Other alternatives, like π-type hydrogen-bonded or slipped parallel stacked arrangements, are energetically definitely unfavorable and, hence, very unlikely to occur.

Acknowledgements

The calculations were performed on the Cluster of Digital Alpha Servers (2100 4/275 and 5/375) of the computer center of the University of Vienna and on local RISC 6000/550 and Silicon Graphics workstations at the Institute of Theoretical Chemistry and Radiation Chemistry of the University of Vienna. The author is grateful for ample supply of computer time on these installations.

References

[1] Maroncelli M, Hopkins GA, Nibler JW, Dyke TR (1985) J Chem Phys **83**: 2129
[2] Hopkins GA, Maroncelli M, Nibler JW, Dyke TR (1985) Chem Phys Lett **114**: 97
[3] Anex DS, Davidson ER, Douketis C, Ewing GE (1988) J Phys Chem **92**: 2913
[4] Jucks KW, Miller RE (1988) J Chem Phys **88**: 2196
[5] Jucks KW, Miller RE (1988) J Chem Phys **88**: 6059
[6] Miller RE (1988) Science **240**: 447
[7] Ruoff RS, Emilsson T, Klots TD, Chuang C, Gutowsky HS (1988) J Chem Phys **89**: 138
[8] Meyer H, Kerstel ERT, Zhuang D, Scoles G (1989) J Chem Phys **90**: 4623
[9] Kerstel ERT, Lehmann KK, Gambogi JE, Yang X, Scoles G (1993) J Chem Phys **99**: 8559
[10] Kerstel ERT, Scoles G, Yang X (1993) J Chem Phys **99**: 876
[11] Yang X, Kerstel ERT, Scoles G, Bemish RJ, Miller RE (1995) J Chem Phys **103**: 8828
[12] Karpfen A (1996) J Phys Chem **100**: 13474
[13] Karpfen A (1998) J Phys Chem **102A**: 9286
[14] Dykstra CE (1996) J Mol Struct (Theochem) **362**: 1
[15] Cabaleiro-Largo EM, Ríos MA (1998) J Chem Phys **108**: 3598
[16] Frisch MJ, Trucks GW, Schlegel HB, Gill PMW, Johnson BG, Robb MA, Cheeseman JR, Keith TA, Petersson GA, Montgomery JA, Raghavachari K, Al-Laham MA, Zakrzewski VG, Ortiz JV, Foresman JB, Cioslowski J, Stefanov BB, Nanayakkara A, Challacombe M, Peng CY, Ayala PY, Chen W, Wong MW, Andres JL, Replogle ES, Gomperts R, Martin RL, Fox DJ, Binkley JS, Defrees DJ, Baker J, Stewart JJP, Head-Gordon M, Gonzalez C, Pople JA (1995) Gaussian 94, Revision C 2 Gaussian Inc, Pittsburgh, PA
[17] Møller C, Plesset MS (1934) Phys Rev **46**: 618
[18] Karpfen A (1983) Chem Phys **79**: 211
[19] Kofranek M, Karpfen A, Lischka H (1987) Chem Phys **113**: 53
[20] Kofranek M, Lischka H, Karpfen A (1987) Mol Phys **61**: 519
[21] Kurnig IJ, Lischka H, Karpfen A (1990) J Chem Phys **92**: 2469
[22] Ditchfield R, Hehre WJ, Pople JA (1971) J Chem Phys **54**: 724
[23] Hehre WJ, Ditchfield R, Pople JA (1972) J Chem Phys **56**: 2257
[24] Huzinaga S (1965) J Chem Phys **42**: 1293
[25] Huzinaga S (1971) Approximate Atomic Functions I. University of Alberta, Edmonton, Canada
[26] Boys SF, Bernardi F (1970) Mol Phys **19**: 553
[27] Botschwina P, Heyl Ä, Oswald M, Hirano T (1997) Spectrochim Acta **A53**: 1079
[28] Botschwina P, Horn M, Seeger S, Flügge J (1993) Mol Phys **78**: 191
[29] Botschwina P, Horn M, Matuschewski M, Schick E, Sebald P (1997) J Mol Struct (Theochem) **400**: 119
[30] Lafferty WJ, Lovas F (1978) J Phys Chem Ref Data **7**: 441
[31] Fowler PW, Diercksen GHF (1990) Chem Phys Lett **167**: 105

Received November 30, 1998. Accepted (revised) December 21, 1998

Proton Motion and Proton Transfer in the Formic Acid Dimer and in 5,8-Dihydroxy-1,4-naphthoquinone: A PAW Molecular Dynamics Study

Katharina Wolf, Alexandra Simperler, and **Werner Mikenda**[*]

Institut für Organische Chemie, Universität Wien, A-1090 Vienna, Austria

Summary. A molecular dynamics study on proton motion and (double) proton transfer in the formic acid dimer (*FAD*) and in 5,8-dihydroxy-1,4-naphthoquinone (*DHN*) is reported that has been performed with the Projector Augmented Wave method (PAW). PAW trajectories were calculated with a time interval of 0.12 fs, for evolution time periods up to 20 ps, and for temperatures in the range 500–700 K. Two basic situations can be clearly distinguished: normal periods that correspond to normal asymmetric O–H$\cdots$O hydrogen bonds, where the proton remains trapped at one oxygen atom, and active periods that correspond to (near-)symmetric O$\cdots$H$\cdots$O hydrogen bonds, where the proton undergoes large amplitude motions between the two adjacent oxygen atoms. Within the active periods one may distinguish between isolated transitions, where a proton just moves from one to the other oxygen atom, crossing-recrossing events, where a proton moves from one to the other oxygen atom but almost immediately turns back, and shuttling periods, where a proton undergoes several consecutive transitions. Moreover, one may also distinguish between single processes, where only one O–H$\cdots$O group is involved, and double processes, where both O–H$\cdots$O groups are simultaneously involved. It is shown that a reasonable and descriptive understanding of the active processes can be obtained by considering the time evolution of the potential energy that governs the motion of the proton between the two adjacent oxygen atoms. A main difference between *FAD* and *DHN* concerns the double proton transfer processes. In the first case these are almost exclusively simultaneous one-step processes, whereas in the second case these are mainly two-step processes, *i.e.* two successive single transitions. This difference can be attributed to the fact that with *DHN* a single proton transfer process yields the metastable 4,8-dihydroxy-1,5-naphthoquinone tautomer, whereas with *FAD* single proton transfer does not result in a metastable intermediate.

Keywords. 5,8-Dihydroxy-1,4-naphthoquinone; Formic acid dimer; Molecular dynamics; Projector Augmented Wave method; Proton transfer.

Protonentransfer im Dimer der Ameisensäure und in 5,8-Dihydroxy-1,4-naphthochinon: Eine PAW-Molekulardynamik-Studie

Zusammenfassung. Mit Hilfe der *Projector Augmented Wave*-Methode (PAW) wurden Molekulardynamiksimulationen über die Protonenbewegung und den Protonentransfer im Dimer der

---

[*] Corresponding author

Ameisensäure (*FAD*) und in 5,8-Dihydroxy-1,4-naphthochinon (*DHN*) durchgeführt. PAW-Trajektorien wurden mit einem konstanten Zeitintervall von 0.12 fs, über Zeiträume von bis zu 20 ps und für Temperaturen im Bereich 500–700 K berechnet. Zunächst lassen sich zwei grundlegende Situationen unterscheiden: normale Perioden, die einer normalen asymmetrischen O–H·· O-Wasserstoffbrückenbindung entsprechen, bei der das Proton eindeutig an ein Sauerstoffatom gebunden ist, und aktive Perioden, die einer annähernd symmetrischen O·· H·· O-Wasserstoffbrückenbindung entsprechen, bei denen sich das Proton mit deutlich größerer Amplitude zwischen beiden benachbarten Sauerstoffatomen bewegt. Weiters kann man innerhalb der aktiven Perioden zwischen isolierten Übergängen, bei denen das Proton von einem zum anderen Sauerstoff wechselt, *crossing-recrossing*-Prozessen, bei denen sich das Proton ebenfalls von einem zum anderen Sauerstoff bewegt, aber sofort wieder zurückkehrt, und *shuttling*-Perioden, bei denen das Proton mehrere aufeinanderfolgende Übergänge ausführt, unterscheiden. Darüber hinaus läßt sich noch zwischen Einzelprozessen, an denen nur eine O–H·· O Gruppe beteiligt ist, und Doppelprozessen, an denen beide O–H·· O Gruppen gleichzeitig beteiligt sind, differenzieren. Eine physikalisch sinnvolle und anschauliche Erklärung der aktiven Prozesse erreicht man durch die Betrachtung der zeitlichen Entwicklung der potentiellen Energie, die die Bewegung des Protons zwischen den beiden Sauerstoffatomen bestimmt. Soweit Doppel-Protonentransfer-Prozesse betroffen sind, besteht zwischen *FAD* und *DHN* ein auffallender Unterschied. Während diese im ersten Fall fast ausschließlich simultane Einstufenprozesse sind, findet man im zweiten Fall überwiegend Zweistufenprozesse, das heißt, es handelt sich um zwei aufeinanderfolgende Einzelübergänge. Dies läßt sich dadurch erklären, daß bei *DHN* ein einfacher Protonentransfer zu einem metastabilen Zwischenzustand führt (5,8-Dihydroxy-1,4-naphtochinon), was bei *FAD* nicht der Fall ist.

Introduction

Recently, we have reported a Projector Augmented Wave (PAW) [1] molecular dynamics study on proton motion and intramolecular proton transfer in malonaldehyde [2, 3]. PAW is based on the direct molecular dynamics approach of *Car* and *Parrinello* [4], which combines classical dynamics with quantum mechanical forces and which yields full molecular dynamics at finite temperatures on a picosecond time scale. Basically, since all nuclear motions are treated classically, no quantum effects such as proton tunnelling or zero point motion are taken into account. Hence, the PAW calculations establish a high temperature approach for dynamic processes, where quantum phenomena are negligible or at least less important (whereas common quantum mechanical studies that focus on tunnelling effects rather correspond to a low temperature limit approximation). Within the framework of the PAW approach, proton motion in malonaldehyde can reasonably well be understood in terms of the full dynamics of the molecule: at each moment (time step), the potential which governs the proton motion between the two adjacent oxygen atoms, is determined by the current molecular geometry. Three typical situations could be distinguished: (*i*) normal periods, in which the proton remains firmly trapped at one oxygen atom, (*ii*) isolated transitions, where the proton rapidly moves from one to the other oxygen atom, and (*iii*) shuttling transition regions, where several consecutive proton transitions take place.

In the present paper we report PAW molecular dynamics studies on two more complex systems that are capable of double proton transfer processes: the cyclic formic acid dimer (*FAD*) and 5,8-dihydroxy-1,4-naphthoquinone (naphtazarine, *DHN*). *FAD* (Fig. 1) belongs to the most extensively studied model systems for the investigation of hydrogen bonded dimers and of intermolecular double proton

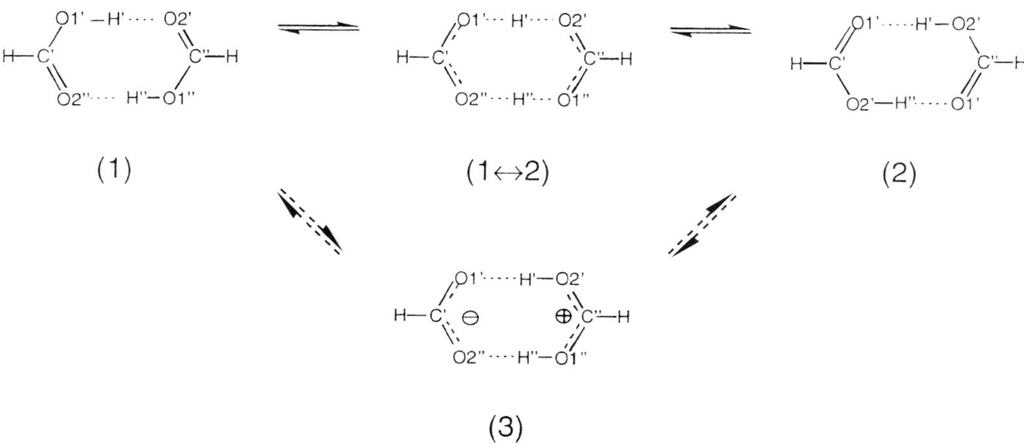

Fig. 1. Prototropic isomers and double proton transfer transition state of the cyclic
formic acid dimer (*FAD*)

transfer processes which play an important role in various fields of chemistry and
biochemistry. Structural and spectroscopic properties of *FAD* have been the subject
of extensive experimental [5–13] and theoretical [14–24] work. Nevertheless,
because of problems with the rather complex and poorly resolved spectra,
tunnelling splitting in *FAD* has never been observed experimentally so far.
According to theoretical studies, double proton transfer in *FAD* most likely takes
place by a simultaneous one-step mechanism *via* a D_{2h} symmetric transition state
$((1) \rightarrow (1 \leftrightarrow 2) \rightarrow (2)$ in Fig. 1) [25–31].

For *DHN* (Fig. 2), which has been chosen as an example for intramolecular
double proton transfer systems, only minor experimental [32–40] and theoretical
[41–45] studies are available in the literature. According to IR and laser-induced
fluorescence low-temperature matrix isolation investigations [36, 37], the C_{2v}
symmetric 5,8-dihydroxy-1,4-naphthoquinone (5,8-*DHN*; (*1*) and (*2*) in Fig. 2)
represents the most stable configuration. Recently, the possible isomers of *DHN*
and the transition states for proton transfer reactions (Fig. 2) have been
theoretically studied at the MP2/6-31G*//HF/6-31G level of theory [41].
According to this study, the C_{2h} symmetric 4,8-dihydroxy-1,5-naphthoquinone
(4,8-*DHN*; (*3*) in Fig. 2) is also a (local) energetic minimum ($\Delta E = 37$ kJ · mol^{-1}
relative to 5,8-*DHN*), whereas the most probable transition states for a single and
for a simultaneous double proton transition with C_s and D_{2h} symmetries,
respectively, are saddle points ($\Delta E = 38$ and 53 kJ · mol^{-1}). Based on these results,
the authors claimed that double proton transfer processes in *DHN* should rather
take place by a two-step mechanism involving the metastable 4,8-*DHN* isomer
$((1) \rightarrow (1 \leftrightarrow 3) \rightarrow (3) \rightarrow (2 \leftrightarrow 3) \rightarrow (2)$ in Fig. 2).

In the following sections we give some computational details, present $R(OH)$
time evolutions to visualize the results of the PAW calculations and to distinguish
between different kinds of proton motion events, show potential energy time
evolutions that provide a basis for an understanding of the physical backgrounds
of the proton transfer processes, and finally discuss similarities and differences
between the two title compounds.

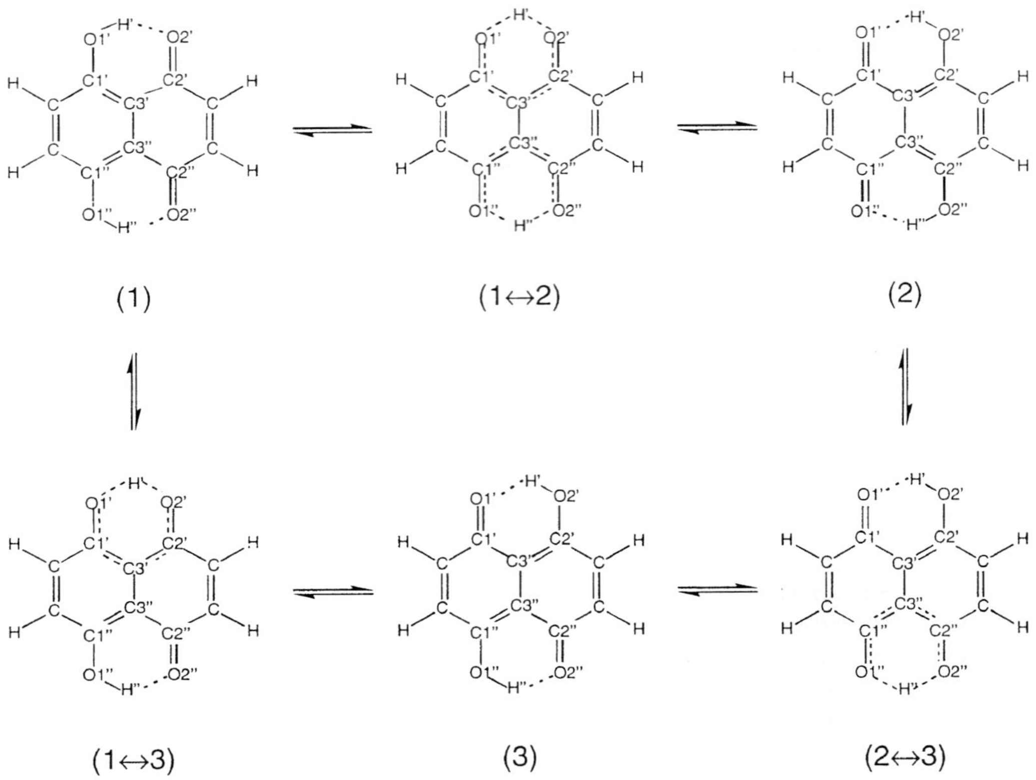

(1)   (1↔2)   (2)

(1↔3)   (3)   (2↔3)

Fig. 2. Prototropic isomers and proton transfer transition states of dihydroxy-naphthoquinone (*DHN*)

Methods

The PAW molecular dynamics simulations were performed for evolution time periods up to 20 ps with constant time intervals of 0.1209 fs (5 a.u.). The temperature of the molecular dynamics runs (between 500 K and 800 K) was controlled with the *Nosè-Hoover* thermostat [46, 47]. *Perdew* and *Zunger's* [48] parametrization of the density functional, based on the results of *Ceperley* and *Alder* [49], was used, and the generalized gradient corrections of *Becke* [50] and *Perdew* [51] were applied. The basis set included plane waves with a cutoff of 30 Ry ($= 39.4\,\mathrm{kJ \cdot mol^{-1}}$); the electron density was represented with a cutoff of 60 Ry. The plane waves were augmented with s-type projector functions for hydrogen atoms and with s-, p-, and d-type projectors for carbon and oxygen atoms. For a comparison of the performance of the PAW calculations with other more common quantum chemical methods, geometry optimizations were performed with the Gaussian92 [52] programs at several levels of theory using the 6-31G(d,p) basis set throughout. Selected geometric and energetic data are compiled in Tables 1 and 2 along with the corresponding zero temperature PAW values. Tables 1 and 2 indicate that the hydrogen bond strengths may be slightly overestimated by the PAW calculations and that our proton transfer barrier heights may be somewhat too low. Since we are exclusively dealing with qualitative features and with a qualitative picture of the proton transfer processes at finite temperatures, the accuracy should be sufficient. What is more, we find a very good agreement between the experimental and the PAW calculated $\nu(\mathrm{OH})$ frequencies (see below) which gives strong evidence that our force fields should indeed be realistic.

Table 1. Selected bond distances (pm) and angles ($°$), and proton transfer barriers ΔE(kJ $\cdot$ mol^{-1}) of *FAD* as obtained from experimental and theoretical data

	exp[a]	HF[b]	MP2[b]	B3LYP[b]	B3P86[b]	PAW
(1)						
O–H	1.036	96.3	99.4	100.7	101.5	106.8
H$\cdots$O	1.667	183.1	171.2	164.4	157.6	143.4
O$\cdots$O	2.703	278.9	270.5	265.0	259.1	250.2
O–H$\cdots$O	180	174	179	179	180	180
C–O	1.323	129.8	132.0	131.0	130.1	129.9
C=O	1.220	119.6	123.0	122.6	122.7	124.0
C$\cdots$C		340.3	383.5	377.7	371.5	364.5
(1$\leftrightarrow$2)						
H$\cdots$O		118.8	120.5	121.0	120.4	122.5
O$\cdots$O		237.6	240.9	241.9	240.8	244.9
O–H$\cdots$O		179	178	179	178	177
C$\cdots$O		124.2	126.9	126.5	126.1	127.0
C$\cdots$C		349.9	353.7	354.6	353.3	359.0
$\Delta E = E(1\leftrightarrow2) - E(1)$	80.1[c]	69.6	34.3	22.5	15.6	8.9

[a] Geometric data from electron diffraction measurements[13]; [b] 6–31G(d,p) basis set; [c] estimated from NIR measurements of ν(O–H) overtone bands [12]

Table 2. Selected bond distances (pm) and angles ($°$), and proton transfer barriers ΔE(kJ $\cdot$ mol^{-1}) of *DHN* as obtained from theoretical data

	HF[a]	MP2[a]	B3LYP[a]	B3P86[a]	PAW
(1)					
O–H	95.2	98.7	0.99.6	100.1	102.9
H$\cdots$O	184.3	172.0	1.67.7	1.63.1	167.1
O$\cdots$O	264.6	260.8	2.57.7	254.6	261.3
O–H$\cdots$O	140	148	148	150	151
C–O	133.0	134.5	1.33.6	132.7	134.2
C=O	120.7	125.4	1.24.9	124.7	126.6
C$\cdots$C	247.3	248.2	2.47.7	246.4	249.8
(3)					
O–H	95.9	101.0	101.8	103.0	104.5
H$\cdots$O	179.7	159.9	157.9	151.6	163.1
O$\cdots$O	261.9	253.3	252.1	248.0	259.9
O–H$\cdots$O	142	152	151	153	152
C–O	131.4	132.8	132.2	131.3	133.6
C=O	121.7	126.7	126.1	126.0	127.1
C$\cdots$C	244.2	246.1	245.9	244.6	249.2
$\Delta E = E(3) - E(1)$	43.8	29.5	19.1	16.5	13.1
$\Delta E = E(1\leftrightarrow3) - E(1)$	67.6	31.6	21.5	16.9	16.0
$\Delta E = E(1\leftrightarrow2) - E(1)$	125.5	49.0	35.1	26.3	20.0

[a] 6–31G(d,p) basis set

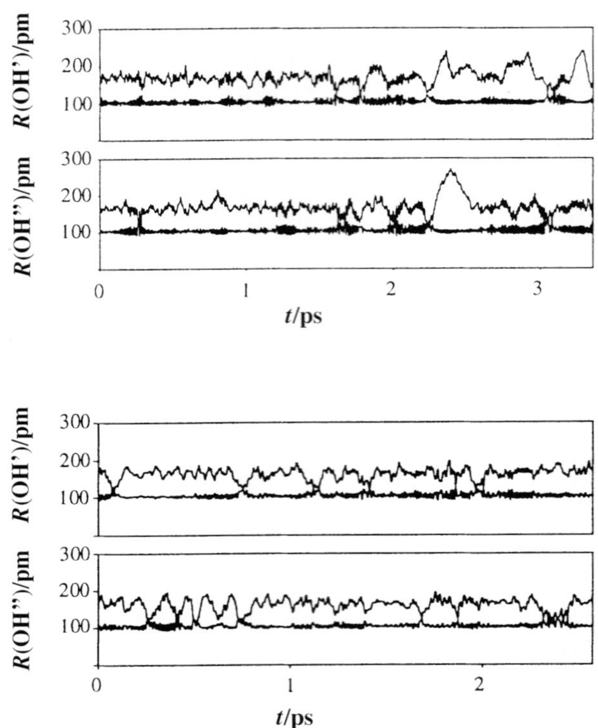

Fig. 3. Time evolutions of the R(OH) distances of *FAD* (top) and *DHN* (bottom)

Results and Discussion

PAW trajectories

In Fig. 3, selected cuts of time evolutions of the two R(OH$'$) and the two R(OH$''$) distances are shown for *FAD* (time period: 3.4 ps, temperature: 700 K) and for *DHN* (time period: 2.7 ps, temperature: 500 K). At first glance one can clearly distinguish between two fundamental situations: normal periods, where the two R(OH) trajectories are well separated from each other (see also Fig. 4a), and active periods, where the two R(OH) trajectories show one or more cross over points (see also Figs. 4b–h). Within the normal periods the proton remains trapped at one oxygen atom and undergoes a stationary motion that corresponds to the ν(OH) stretching mode of a normal, clearly asymmetric O–H$\cdot\cdot$O hydrogen bond. The vibrational amplitude is typically 100 pm, and the average frequencies are 80 THz and 84 THz (2800 and 2700 cm^{-1}) for *FAD* and *DHN*, respectively. In contrast, within the active periods the proton undergoes large amplitude motions between the two adjacent oxygen atoms, with an amplitude of typically about 300 pm.

Closer inspection of the trajectories reveals that the processes within the active periods are largely variable, stochastic events. Selected examples are shown in Fig. 4 by blow-ups of the time evolutions of the R(OH$'$) and R(OH$''$) distances. For a phenomenological description, to a first approximation the active periods can be classified according to two main criteria. First, according to the number of cross-

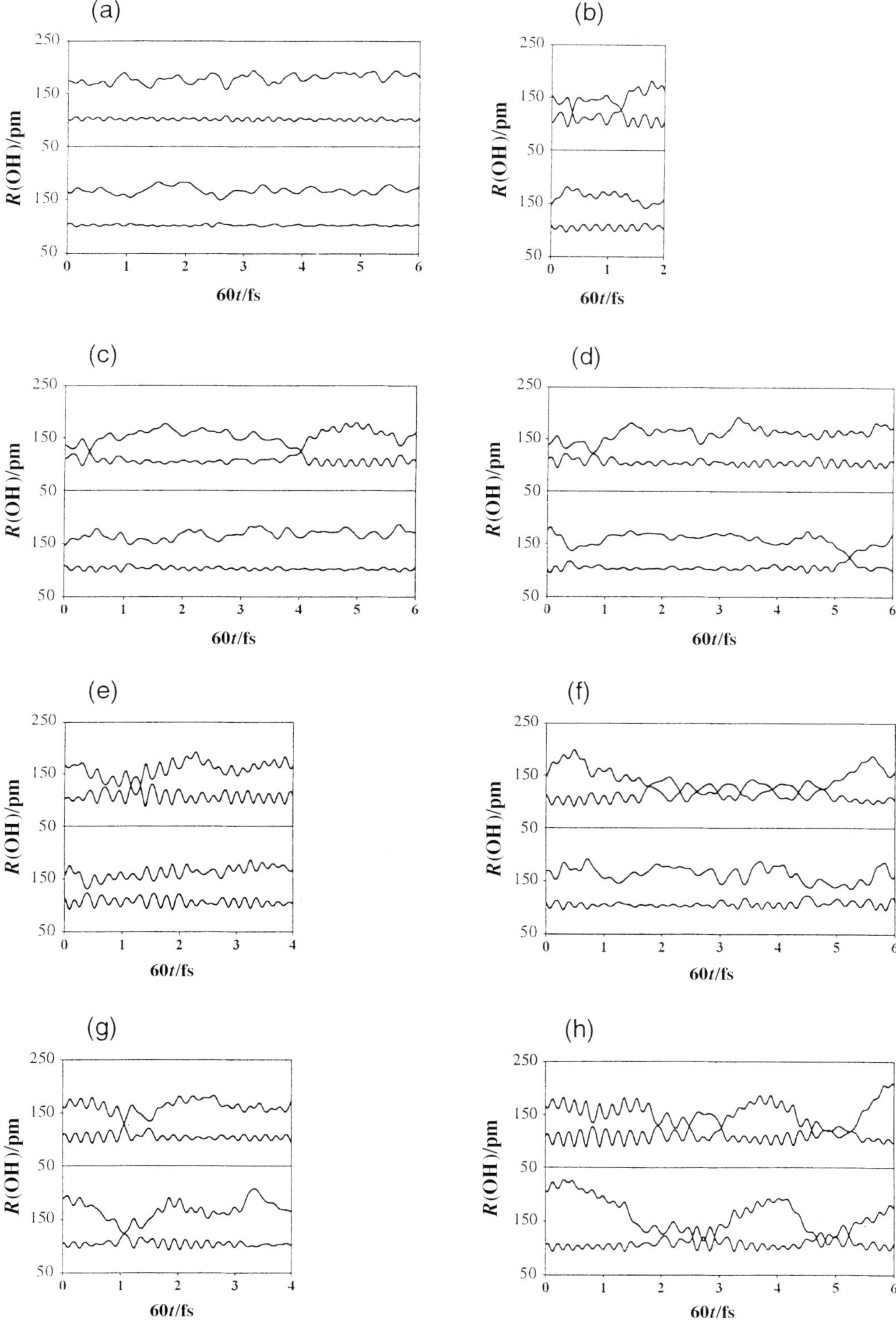

Fig. 4. Time evolutions of the R(OH) distances showing (a) a normal period (*DHN*), (b) a consecutive single proton transfer process (*DHN*), (c) a successive single proton transfer process (*DHN*), (d) a successive double proton transfer process (*DHN*), (e) a single concerted crossing-recrossing event (*DHN*), (f) a single proton shuttling period (*DHN*), (g) a simultaneous double proton transfer (*FAD*), and (h) a double proton shuttling period (*FAD*)

over points we can distinguish between (*i*) isolated proton transfer processes (Figs. 4b–d,g) where the proton just moves from one to the other oxygen atom (one cross over point, $O–H\cdot\cdot O \rightarrow O\cdot\cdot H–O$), (*ii*) crossing-recrossing processes (Fig. 4e) where the proton moves from one to the other oxygen atom but almost immediately goes back to the first oxygen (two cross over points, $O–H\cdot\cdot O \rightarrow O\cdot\cdot H–O \rightarrow O–H\cdot\cdot O$), and (*iii*) shuttling periods (Figs. 4f, h), where the proton undergoes several consecutive transitions (three or more cross over points, $O–H\cdot\cdot O \rightarrow O\cdot\cdot H–O \rightarrow O–H\cdot\cdot O \rightarrow O\cdot\cdot H–O \rightarrow \rightarrow$). Second, according to the number of $O–H\cdot\cdot H$ groups that are simultaneously active we can distinguish between (*i*) single processes (Figs. 4b–f) where only one $O–H\cdot\cdot O$ group is involved and (*ii*) double processes (Figs. 4g–h) where both $O–H\cdot\cdot O$ groups are simultaneously involved. To be sure, these classifications are neither unambiguous nor do they account for the full variety of the observed processes, but they provide a reasonable basis for the following discussions.

Geometric considerations

In a foregoing study on malonaldehyde [2–3] it has been shown that from a geometrical point of view normal and active periods are mainly characterized by long and short $R(O\cdot\cdot\cdot O)$ distances, respectively, although there exists no well defined borderline which could serve as a necessary or a sufficient criterion for a clear cut distinction. The very same is true for *FAD* and *DHN*: the average $R(O\cdot\cdot\cdot O)$ distances within the normal regions amount to 272 pm and 263 pm, whereas the average $R(O\cdot\cdot\cdot O)$ distances of the cross-over points are 248 pm and 239 pm, respectively. From an analysis of the geometric data by statistical methods, several additional geometric parameters could also be determined which on average show more or less systematic differences between normal and active periods. In summary, however, it is not possible to reliably discriminate between normal periods and (cross-over points of) active periods by purely geometric arguments, *i.e.* there are no definite requirements for activity to take place, and the individual active periods may be associated with largely different geometries.

Energetic considerations

Previously it has been shown [2–3] that a reasonable and descriptive understanding of the driving forces that govern the motion of the proton between the two adjacent oxygen atoms can be obtained by considering potential energy time evolutions ($E(\rho,\,t)$). Some selected examples are shown in Figs. 5–7 by staggered plots of consecutive ($\Delta t = 2.4$ fs) potentials $E(\rho)_t$. The single potentials $E(\rho)_t$ (*i.e.* the single time frames) were calculated by taking the corresponding PAW geometries, fixing all atoms (except the proton under consideration), moving the proton stepwise along a proper proton transfer reaction coordinate, $\rho = (R(O1H)\cdot\cos\theta(O2O1H))/R(O1O2)$, and optimizing the proton position at each step. The potential $E(\rho)_t$ obtained by such a point to point calculation represents the energetic precondition for a hypothetical, infinitely rapid proton transfer for the molecular geometry just given at that time step. Finally, the complete potential energy time evolution $E(\rho,t)$ (*i.e.* the staggered plots) illustrates the energetic situation as experienced by the

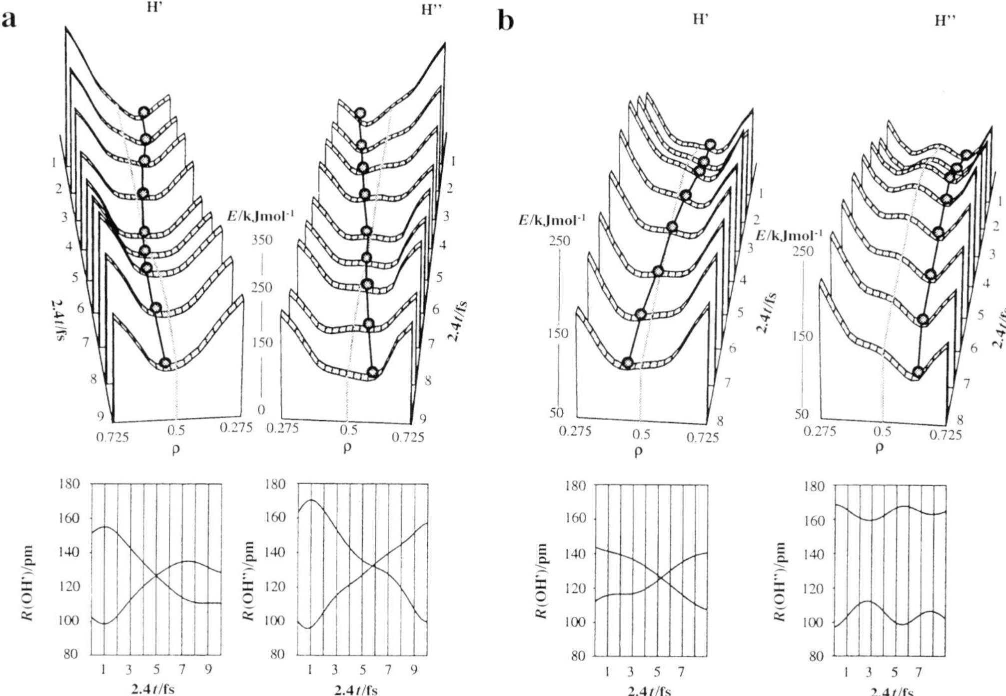

Fig. 5. Time evolutions of $E(\rho)$ potentials (time interval between consecutive frames: 2.42 fs; the circles indicate the actual proton positions) and corresponding time evolutions of the $R(\mathrm{OH})$ distances for (a) an isolated double proton transfer process (*FAD*) and for (b) an isolated single proton transfer process (*DHN*)

proton on the fly which permanently changes due to the dynamically changing molecular geometry. In Figs. 5–7, such potential energy time evolutions $E(\rho,t)$ are shown by pairs of staggered plots ($\Delta t = 2.4$ fs) along with the actual proton motions as obtained from the PAW trajectories (indicated by circles). Additionally, the time evolutions of the $R(\mathrm{OH})$ distances are shown; each vertical line corresponds to the equally numbered time frame of the plot.

Figures 5a and 5b show nine and eight snapshots (total time periods: 19.2 and 16.8 fs) of an isolated double proton transfer process in *FAD* and of an isolated single proton transfer process in *DHN*, respectively. The time evolutions $E(\rho, t)$ for the active protons (H$'$ and H$''$ for *FAD*, H$'$ for *DHN*) start with normal, strongly asymmetric potentials, whose minima are located at one oxygen atom; then they change to broad, (near-)symmetric single-minimum or shallow double-minimum potentials, and end with again strongly asymmetric potentials with the minima now being located at the other oxygen atom. As to the protons, at the beginning they are firmly attached to one oxygen atom; then, following the $E(\rho, t)$ gradient, they rapidly move towards the other oxygen, where they finally remain trapped.

Figure 6a shows 21 snapshots (48 fs) of a double proton shuttling period in *FAD*, Fig. 6b shows 15 snapshots (33.6 fs) of a single proton shuttling period in *DHN*. Like isolated processes, shuttling periods are also associated with broad, (near-)symmetric single-minimum or shallow double-minimum potentials. The

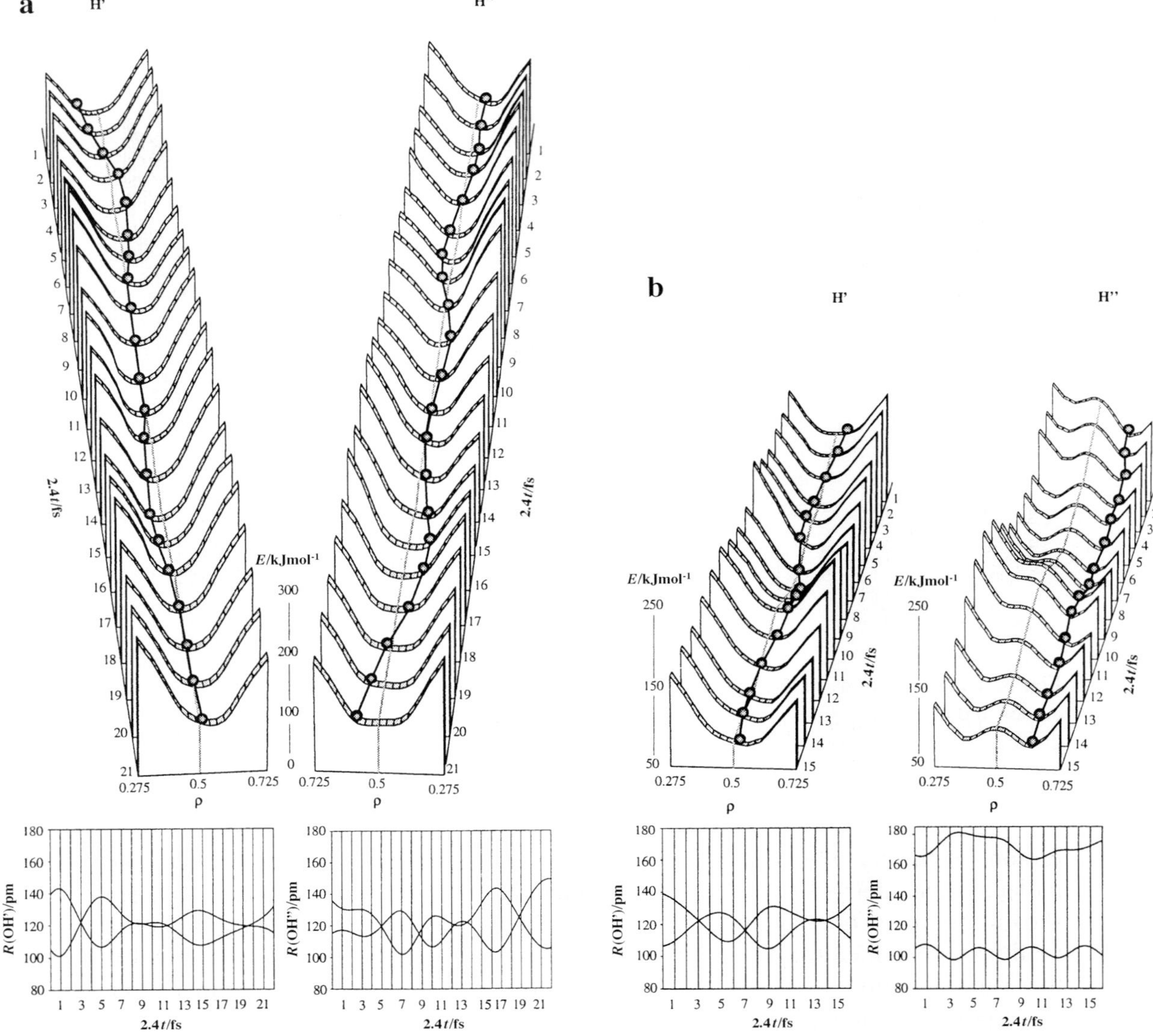

Fig. 6. Time evolutions of $E(\rho)$ potentials (time interval between consecutive frames: 2.42 fs; the circles indicate the actual proton positions) and corresponding time evolutions of the $R(OH)$ distances for (a) a double proton shuttling period (*FAD*) and for (b) a single proton shuttling period (*DHN*)

main difference is that in the first case the normal asymmetric potential is readily re-established, whereas in the latter case the abnormal (near-)symmetric potential persists for a longer time period, thus leading to the observed large-amplitude proton motions. In Fig. 7, as a final example, an outstanding crossing-recrossing process in *FAD* is shown (10 snapshots, 21.6 fs) where the potential remains clearly asymmetric during the whole period. In this case the observed large amplitude proton motion obviously results from an accidental, exceptionally large kinetic energy of the proton, as already indicated by the abnormal large elongation in

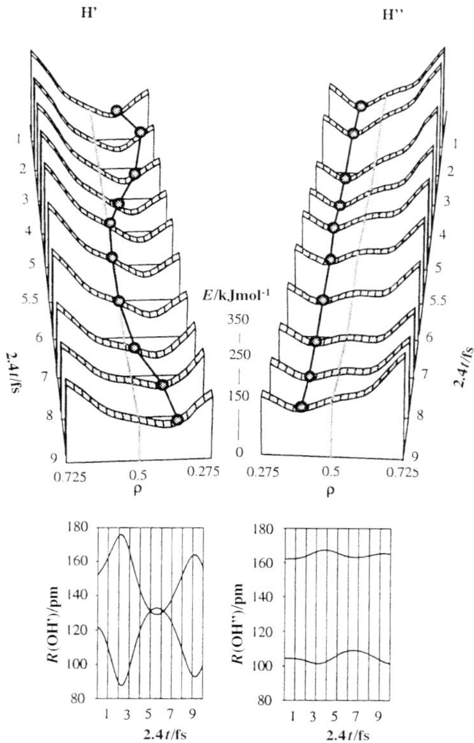

Fig. 7. Time evolutions of $E(\rho)$ potentials (time interval between consecutive frames: 2.42 fs; the circles indicate the actual proton positions) and corresponding time evolutions of the $R(OH)$ distances for a crossing-recrossing process (*FAD*)

frame 2. This clearly shows that the proton motion is not only governed by the potential $E(\rho, t)$, but also by kinetic factors, *i.e.* by the proton's momentum.

From the above discussion, the physical backgrounds of the proton motion, as obtained by the PAW calculations, may be summarized as follows. (*i*) At any given time step, the potential energy $E(\rho)_t$ that governs the proton motion between the two adjacent oxygen atoms is determined by the current molecular geometry, which on its part is determined by the full dynamics of the molecule. Basically, the proton motion is not rigorously determined, but (only) governed by the potential $E(\rho)_t$, since the actual motion also depends on the current kinetic energy (the momentum) of the proton. (*ii*) On average, the normal periods correspond to normal, clearly asymmetric O–H · · O hydrogen bonds, *i.e.* the average geometries are similar to those of the stable isomers of the two compounds – (*1*) and (*2*) – and $E(\rho)_t$ is an asymmetric potential with the minimum clearly located at one of the two oxygen atoms in which the proton undergoes a normal ν(O–H) stretching vibration. (*iii*) On the other hand, on average the active periods correspond to (near-)symmetric O · · H · · O hydrogen bonds, *i.e.* the average geometries are similar to those of the proton transfer transition states of the two compounds – (*1*↔*2*), (*1*↔*3*), and (*2*↔*3*) – and $E(\rho)_t$ is a broad, (near-)symmetric single- or double-minimum potential in which the proton can undergo large amplitude motions between the two adjacent oxygen atoms. This motion rather corresponds to a quasi-stationary $\nu(O \cdots H \cdots O)$ stretching vibration than to consecutive single proton transfers.

FAD versus DHN

Based on the above classification and on the physical backgrounds of active processes, we finally focus on similarities and/or differences between the two title compounds. With *FAD* we mainly find two different situations (about 90%), isolated double proton transfer processes and single crossing-recrossing processes, whereas double and single proton shuttling periods are rather scarce (about 10%) and single proton transfer processes are totally absent. For the double proton transfer processes (Fig. 4g), the delay between the two proton motions, as measured by the difference between the two cross-over points, ranges from 0.1 to 11 fs, which gives justification to talk about simultaneous or concerted one-step processes (note that 12 fs corresponds to a normal $\nu(OH)$ vibrational cycle). Similarly, the crossing-recrossing events (Fig. 4e) are also concerted one-step processes; the differences between the two cross over points are less than 12 fs in all instances.

With *DHN* the situation is largely different and distinctly more complex. First of all, simultaneous processes, *i.e.* double proton transfers and double proton shuttling periods, are rather scarce ($< 15\%$). Second, similar to *FAD* single proton crossing-recrossing (Figs. 4b, c) comprises about 40% of the active periods; however, the differences between the two cross-over points range typically from 10 to 30 fs, *i.e.* most of them are distinctly larger than with *FAD*, and in an extreme case we found a delay of as much as 220 fs. This means that with *DHN* crossing-recrossing not only takes place as clear cut concerted one-step process, as is the case with *FAD*, but also as successive two-step process (with many additional cases in-between). Consistently, we also find successive two-step double proton transfer processes ($> 10\%$); they start with an isolated single proton transition at one O–H $\cdots$ O group followed by a second isolated proton transition at the other O–H $\cdots$ O group (Fig. 4d). Quite noticeable, the delay between the two transitions is always rather large; it ranges from 30 to 270 fs. Finally, the remaining 25% of the active periods of *DHN* comprise single proton shuttling periods (Fig. 4f).

The observed differences between *FAD* and *DHN* can reasonably well be understood by considering the (zero temperature) energies and stabilities of the prototropic isomers, $(1) = (2)$ and (3), and of the proton transfer transition states, $(1 \leftrightarrow 2)$ and $(1 \leftrightarrow 3) = (2 \leftrightarrow 3)$, (Figs. 1 and 2). For *FAD*, quantum chemical calculations yield the two equivalent minimum energy structures (1) and (2), and the D_{2h} symmetric double proton transfer transition state $(1 \leftrightarrow 2)$ which is a saddle point (Table 1), whereas the single proton transfer product (3) is not a stationary state (*i.e.* is not a local minimum). Consequently and consistently, proton transfer processes (almost) exclusively take place by a concerted one-step mechanism that simultaneously involves both O–H $\cdots$ O groups: $(1) \rightarrow (1 \leftrightarrow 2) \rightarrow (2)$. What is more, single processes that involve only one O–H $\cdots$ O group are almost exclusively restricted to concerted crossing-recrossing events where the proton undergoes a large amplitude vibration for just one cycle, which means that situations which cause single proton activity are highly unstable and persist for very short time periods only.

With *DHN* the situation is distinctly different. Quantum chemical calculations show that besides the two global minima (*1*) and (*2*) the product of a single proton transfer reaction (*3*) is also a minimum on the PES (*i.e.* 4,8-*DHN* is a metastable tautomer), and the transition states for single proton transfer (*1↔3*) and (*2↔3*) are also saddle points (Table 2). What is more, at all levels of theory the energies of the different species increase within the series (*1*) = (*2*) < (*3*) < (*1↔3*) = (*2↔3*) < (*1↔2*). Consequently, single proton transfer processes should be energetically preferred over (simultaneous) double proton transfer processes. Quite consistently, the majority of active processes observed within the PAW trajectories of *DHN* are single processes that are confined to only one O–H· ·O group and involve the single proton transfer transition states (*1↔3*) = (*2↔3*) and the metastable isomer (*3*): one- or two-step crossing-recrossing events, consecutive two-step double proton transfer processes, and single proton shuttling. In outstanding cases the meastable prototropic isomer (*3*) remains stable for more than 200 fs. On the other hand, because of the high energy of the double proton transfer transition state (*1↔2*), concerted double proton transfer or double proton shuttling processes are rather scarce.

Conclusions

The PAW molecular dynamics studies about proton motion and proton transfer in *FAD* and in *DHN* reported in this paper on the one hand largely confirm the results of a foregoing study on malonaldehyde [2, 3]. On the other hand, since the two compounds under consideration are capable of double proton transfer, the present study should further contribute to the understanding of these processes.

At first glance, within the PAW trajectories we can distinguish between normal and active periods. In the former case, a proton remains firmly attached at one oxygen atom, in the latter case a proton undergoes large amplitude motions between the two adjacent oxygen atoms which may (but not necessarily must) result in proton transfer. From a geometric point of view there exist no definite requirements for proton activity to occur, but on the average, within the active periods the $R(O· · ·O)$ distances are distinctly shorter (by about 25 pm) than within the normal periods.

Within the active periods we find isolated transitions (one transition: O–H· ·O→O· ·H–O), crossing-recrossing events (two transitions: O–H· ·O→ O· ·H–O→O–H· ·O), and shuttling periods (several consecutive transitions: O–H· ·O→O· ·H–O→O–H· ·O→O· ·H–O→ →). We also find single and double processes that involve only one or (simultaneously) both O· ·H· ·O groups respectively, and we can also distinguish between concerted one-step processes and successive two-step processes.

A reasonable and descriptive understanding of the driving forces behind the proton motion can be obtained by considering time evolutions of the potential that governs the motion of the proton between the two adjacent oxygen atoms. Normal periods are associated with asymmetric potentials with the minimum clearly located at one of the two oxygen atoms, as they are characteristic for normal, clearly asymmetric O–H$\cdots$O hydrogen bonds, whereas active periods are associated with broad, (near-)symmetric single- or double-minimum potentials, as they are characteristic for (near-)symmetric O$\cdots$H$\cdots$O hydrogen bonds.

An apparent difference between the two title compounds concerns the mechanism of double proton transitions. With *FAD* these are almost exclusively simultaneous one-step processes, whereas with *DHN* these are preferably two-step processes (*i.e.* two successive single proton transitions). This difference can reasonably well be attributed to the fact that single proton transfer in *DHN* yields the metastable 4,8-*DHN*, the energy of which is safely below the double proton transfer transition state, whereas with *FAD* the double proton transfer transition state is the only saddle point, and single proton transfer does not yield a metastable intermediate state.

Acknowledgements

The authors are grateful to Doz. *P. E. Blöchl*, IBM research division, Zürich (Switzerland), and to Dr. *E. Nusterer* and Prof. *K. Schwarz*, Technical University of Vienna (Austria) for valuable help and discussion. The work was supported by the *Jubiläumsfonds der Österreichischen Nationalbank*, Proj. No. 7237. Ample supply of computer facilities (IBM-RISC 6000/550 and Digital Alpha 2100 5/375) by the Computer Centre of the University of Vienna is kindly acknowledged.

References

[1] Blöchl PE (1994) Phys Chem Rev B **50**: 17953
[2] Wolf K, Mikenda W, Nusterer E, Schwarz K, Ulbricht C (1998) Chem Eur J **4**: 1418
[3] Wolf K, Mikenda W, Nusterer E, Schwarz K (1998) J Mol Struct **448**:201
[4] Car R, Parrinello M (1985) Phys Rev Lett **55**: 2471
[5] Mualem R, Sominska E, Kelner V, Gedanken A (1992) J Chem Phys **97**: 8813
[6] Marechal Y (1988) J Mol Struct **189**: 55
[7] Wachs T, Borchardt D, Bauer SH (1987) Spectrochim Acta A **43**: 965
[8] Bertie JE, Michaelian KH, Eysel HH, Hager D (1986) J Chem Phys **85**: 4779
[9] Bertie JE, Michaelian KH (1982) J Chem Phys **76**: 886
[10] Bournay J, Marechal Y (1975) Spectrochim Acta A **31**: 1351
[11] Rothschild WG (1974) J Chem Phys **61**: 3422
[12] Morita H, Nagakura S (1972) J Mol Spec **42**: 536
[13] Almenningen A, Bastiansen O, Motzfeld T (1969) Acta Chem Scand **23**: 2848
[14] Qian W, Krimm S (1998) J Phys Chem A **102**: 659
[15] Qian W, Krimm S (1998) J Phys Chem A **101**: 5825
[16] Wolfs I, Desseyn HO (1996) J Mol Struct (Theochem) **360**: 81
[17] Chojnacki H, Andzelm J, Nguyen DT, Sokalski WA (1995) Computers Chem **19**: 181
[18] Borisenko KB, Bock CW, Hargittai I (1995) J Mol Struct (Theochem) **332**: 161
[19] Chang YT, Yamaguchi Y, Miller WH, Schaefer HF III (1987) J Am Chem Soc **109**: 7245
[20] Mijoule C, Allavena M, Leclercq JM, Bouteiller Y (1986) Chem Phys **109**: 207
[21] Karpfen A (1984) Chem Phys **88**: 415
[22] Robertson GN, Lawrence MC (1981) Chem Phys **62**: 131
[23] Bosi P, Zerbi G, Clementi E (1977) J Chem Phys **66**: 3376
[24] Del Bene JE, Kochenour WL (1976) J Am Chem Soc **98**: 2041
[25] Lim JH, Lee EK, Kim Y (1997) J Phys Chem A **101**: 2233
[26] Jursic B (1997) J Mol Struct (Theochem) **417**: 89

[27] Chojnacki H (1997) Molecular Engineering **7**: 161

[28] Kim Y (1996) J Am Chem Soc **118**: 1522

[29] Shida N, Barbara PF, Almlöf J (1991) J Chem Phys **94**: 3633

[30] Hayashi S, Umemura J, Kato S, Morokuma K (1984) J Phys Chem **88**: 1330

[31] Graf F, Meyer R, Ha TK, Ernst RR (1981) J Chem Phys **75**: 2914

[32] Reynhardt EC (1992) Mol Phys **76**: 525

[33] Paul SO, Schutte CJH, Hendra PJ (1990) Spectrochim Acta A **46**: 323

[34] Olivieri A, Paul IC, Curtin DY (1990) Magn Res Chem **28**: 119

[35] Herbstein FH, Kapon M, Reisner GM, Lehman MS, Kress RB, Wilson RB, Shiau WI, Duesler EN, Paul IC, Curtin DY (1985) Proc R Soc Lond A **399**: 295

[36] Rentzepis PM, Bondybey VE (1984) J Chem Phys **80**: 4727

[37] Bondybey VE, Milton SV, English JH, Rentzepis PM, (1983) Chem Phys Lett **97**: 130

[38] Anoshin AN, Kopteva TS, Mikhailova KV, Shapiro IO, Shigorin DN (1982) Russ J Phys Chem **56**: 1215

[39] Shiau WI, Duesler EN, Paul IC, Curtin DY (1980) J Am Chem Soc **102**: 4546

[40] Bratan S, Strohbusch F (1980) J Mol Struct **61**: 409

[41] Ramondo F, Bencivenni L (1994) Struct Chem **5**: 211

[42] Schutte CJH, Paul SO, Smit R (1993) J Mol Struct **297**: 235

[43] Anoshin AN, Gastilovich EA, Mishenina KA (1983) Russ J Phys Chem **57**: 867

[44] Anoshin AN, Gastilovich EA, Nekrasov VV, Nurmukhametov RN, Shigorin DN (1983) Russ J Phys Chem **57**: 870

[45] De la Vega JR, Busch JH, Schauble JH, Kunze KL, Haggert BE (1982) J Am Chem Soc **104**: 3295

[46] Nosè S (1984) Mol Phys **52**: 255

[47] Hoover WG (1985) Phys Rev A **31**:1965

[48] Perdew JP, Zunger A (1981) Phys Rev B **23**: 5048

[49] Ceperley D, Alder BJ (1980) Phys Rev Lett **45**: 566

[50] Becke AD (1992) J Chem Phys **96**: 2155

[51] Perdew JP (1986) Phys Rev B **33**: 8822

[52] Frisch MJ, Trucks GW, Schlegel HB, Gill PMW, Johnson BG, Wong MW, Foresman JB, Rob MA, Head-Gordon M, Replogle ES, Gomperts R, Andres JL, Raghavahari K, Binkley JS, Gonzales C, Martin RL, Fox DJ, Defrees DJ, Baker J, Stewart JJP, Pople JA (1993) Gaussian 92, Rev G4. Gaussian, Inc, Pittsburgh, PA

Received November 13, 1998. Accepted (revised) January 12, 1999

Correlation of O–H Stretching Frequencies and O–H···O Hydrogen Bond Lengths in Minerals

Eugen Libowitzky

Institut für Mineralogie und Kristallographie, Universität Wien – Geozentrum, A-1090 Vienna, Austria

Summary. A correlation of O–H stretching frequencies (from infrared spectroscopy) with O···O and H···O bond lengths (from structural data) of minerals was established. References on 65 minerals yielded 125 data pairs for the d(O···O)-v correlation; due to rare or inaccurate data on proton positions, only 47 data pairs were used for the d(H···O)-v correlation. The data cover a wide range of wavenumbers from 1000 to 3738 cm^{-1} and O···O distances from 2.44 to 3.5 Å. They originate from silicates, (oxy)hydroxides, carbonates, sulfates, phosphates, and arsenates with OH$^-$, H$_2$O, or even H$_3$O$_2^-$ units forming very strong to very weak H bonds. The correlation function was established in the form v(cm^{-1}) = 3592–304 · 10^9 · exp(-d(O···O)/0.1321), R^2 = 0.96. Because of deviations from ideal straight H bonds, *i.e.* bent or bifurcated geometry, dynamic proton behavior, but also due to factor group splitting and cationic effects, data scatter considerably around the regression line. The trends of previous correlation curves and of theoretical considerations were confirmed.

Keywords. Correlation diagram; Hydrogen bond length; Infrared spectroscopy; Mineral structure; Stretching frequency.

Korrelation von O–H-Streckfrequenzen und O–H···O-Wasserstoffbrückenlängen in Mineralen

Zusammenfassung. Eine Korrelation zwischen O–H-Streckfrequenzen (aus IR-spektroskopischen Messungen) und O···O- sowie H···O-Bindungslängen (aus Strukturdaten) von Mineralen wurde erstellt. Literaturzitate über 65 Minerale lieferten 125 Datenpaare für die d(O···O)-v-Korrelation. Aufgrund seltener oder ungenauer Daten über Wasserstoffpositionen konnten nur 47 Datenpaare für die d(H···O)-v-Korrelation verwendet werden. Die Daten decken einen weiten Wellenzahlbereich von 1000 bis 3738 cm^{-1} und O···O-Bindungslängen von 2.44 bis 3.5 Å ab. Sie entstammen Silikaten, (Oxi)hydroxiden, Karbonaten, Sulfaten, Phosphaten und Arsenaten mit OH$^-$, H$_2$O, oder sogar H$_3$O$_2^-$ Gruppen, welche sehr starke bis sehr schwache Wasserstoffbrücken ausbilden. Die Korrelationsfunktion wurde in der Form v(cm^{-1}) = 3592–304 · 10^9 · exp(-d(O···O)/0.1321), R^2 = 0.96 erstellt. Aufgrund von Abweichungen von idealen gestreckten H-Brücken, d.h. geknickter oder gegabelter Geometrie, dynamischem Verhalten der Protonen, aber auch wegen Faktorgruppenaufspaltung und Kationeneffekten, streuen die Daten beachtlich um die Regressionslinie. Die Trends früherer Korrelationskurven und theoretischer Berechnungen wurden bestätigt.

Introduction

Hydrogen bonds are usually classified according to bond length, *i.e.* $d(O\cdots O)$, $d(H\cdots O)$, or even $d(O-H)$, and bond strength. The latter is conveniently expressed by the fundamental stretching frequency v of the O–H bond which is easily obtained by infrared (IR) and *Raman* spectroscopy. Thus, very strong H bonds are observed at wavenumbers below $1600\,cm^{-1}$ and at $d(O\cdots O) < 2.50\,\text{Å}$, strong H bonds are characterized by stretching frequencies between 1600 and $3200\,cm^{-1}$ and $O\cdots O$ distances between 2.50 and $2.70\,\text{Å}$, and weak H bonds occur above $3200\,cm^{-1}$ (up to $\sim 3700\,cm^{-1}$) and at $d(O\cdots O) > 2.70\,\text{Å}$ (with a barely defined upper limit beyond $3\,\text{Å}$) [1].

The correlation of decreasing O–H stretching frequency with enhanced H bonding has early been recognized [2] and attributed to the attractive force of the H bond acceptor which shifts the proton off the donor atom and thus attenuates the O–H bond strength. The relation between frequency shift and H bond lengths has been approached by theoretical calculations, (*e.g.* Ref. [3]) and by comparison of experimentally determined structural and spectroscopic data. One of these early correlations [4] used data with wavenumbers between 1780 and $3700\,cm^{-1}$ and $O\cdots O$ distances between 2.44 and $3.36\,\text{Å}$. At that time the authors assumed a frequency of $3700\,cm^{-1}$ for absence of H bonding, whereas the value for the free OH^- ion accepted today is $\sim 3560\,cm^{-1}$ [5]. A later correlation [6] used a similar data set, typically representing a mixture of organic and inorganic compounds. At this time, however, the extremely broad and low-frequent stretching bands of very strong H bonds had already been accepted and they extended the range of the diagram down to $700\,cm^{-1}$ at $d(O\cdots O) = 2.44\,\text{Å}$. Whereas the two latter correlations used a wide range of data values at the expense of reduced precision, a recent correlation on solid hydrates [7] provides good precision in a limited frequency range between 2900 and $3600\,cm^{-1}$ and a distance range of $d(O\cdots O) = 2.60-3.10\,\text{Å}$ and $d(H\cdots O) = 1.65-2.15\,\text{Å}$. In addition, it provides also data for hydrogen bonds with non-oxygen acceptor atoms. The use of selected data (exclusively hydrates, spectra from isotopically diluted samples, only ideal straight H bonds) provides limited scatter and thus it may be used for predictive purposes, assumed that the unknown samples obey to the same quality criteria.

It is the aim of the present paper to present a distance-frequency correlation which provides a wide range of data in terms of $d(O\cdots O)$ or $d(H\cdots O)$ and wavenumber, and which takes data exclusively from inorganic minerals composed of rather abundant elements in nature. It was the intention of the author not to restrict the dataset to a specific group of minerals or to a specific H bond environment. Even if the scatter of data might limit its use for highly accurate predictions, the data should reflect the various deviations from the pure $d-v$ relation which may be encountered in natural solids.

In general, distance-frequency correlations provide important information in cases where bulk crystallographic methods, *e.g.* X-ray and neutron diffraction, are not able to locate protons or $O-H\cdots O$ hydrogen bonds. Common examples are trace hydrogen defects [8] and dynamic behavior or disorder of protons in minerals [9, 10]. For the former case only spectroscopic methods allow to determine the distorted environment (*e.g.* distances) around the structural OH defect, for the latter only the high time resolution of spectroscopic techniques reveals information on the actual proton positions and distances.

Data Selection

In most cases, structural data of minerals were retrieved from the "Inorganic Crystal Structure Database – ICSD 98/1" [11] and taken from the references therein. Only for a few minerals a single reference was available. For the majority of compounds a selection among many structure refinements had to be made resulting in a compromise between the most recent reference, the best refined data (using R values), or the experiment with the best H atom localization (usually from neutron diffraction). In many cases only the latter provided data with sufficient accuracy for a $d(H\cdots O)$ calculation, and only the latter gave accurate angles to classify H bonds into straight or bent ones (with $<O–H\cdots O = 150°$ as the critical case). As a rule, H$\cdots$O distances were rejected if the respective O–H distances were (apparently) shorter than 0.90 Å. In general, room temperature data were used, except for cases with dynamic proton disorder, *e.g.* lawsonite [9] and hemimorphite [12], where the room temperature structure refinements yielded only average disordered positions, whereas the low temperature refinements of the ordered structures gave correct atom sites.

IR spectroscopic data were predominantly retrieved from spectra collections of minerals, *e.g.* Refs [13–15], the most recent one being available also in electronic data format [16]. Transmission spectra of mostly KBr powder pellets were used. In a few cases also single crystal data from oriented sections measured with polarized IR radiation were available. The latter facilitate accurate band assignments even in cases of multiple H bonds using the ratios of polarized band intensities in reconciliation with O–H vector orientations in the structure [17]. In general, in systems with more than one H bond, bands were assigned from low to high wavenumbers to short to long H bond distances, even in cases where a different assignment according to cationic effects had been proposed, *e.g.* malachite [18]. Complicated systems, *e.g.* chalcanthite-$CuSO_4 \cdot 5H_2O$, which display a broad absorption band that cannot be separated into single bands due to a large number of very similar H bonds and vibrational coupling in undeuterated samples, were not included in the correlation.

According to the above mentioned criteria, a number of 65 minerals was finally chosen for the correlation. Mineral names, formulae, structural characteristics, and both structural and spectroscopic references are summed up in Table 1. As far as possible, widely known species were used. Only in a few cases (to cover a rather underrepresented part of the diagram or to set data points of very well studied samples) rare species were included. In cases where minerals display wide and complicated solid solution series, *e.g.* tourmaline, samples close to a well defined endmember composition were used. However, for vesuvianite the crystal chemical environment around the H atoms appeared too variable and complicated to be useful for a reliable correlation.

The samples belong to 5 large groups of compounds: silicates (39 samples, 64 data points), (oxy)hydroxides (8, 12), carbonates (3, 6), sulfates (8, 26), phosphates, and arsenates (7, 17). They contain hydrous species in the form of

Table 1. Mineral names, formulae, structural characteristics, and references for structural and spectroscopic data

Mineral	Formula	Structural elements	Structure	IR data
Mozartite	$CaMnOSiO_3(OH)$	$[SiO_3(OH)]$	[19]	[19]
Topaz	$Al_2(F,OH)_2SiO_4$	$[SiO_4]$	[20]	[21]
Hydrogrossular	$Ca_3Al_2(O_4H_4)_x(SiO_4)_{3-x}$	$[SiO_4]$, $[O_4H_4]$	[22]	[23]
Henritermierite	$Ca_3Mn_2(O_4H_4)(SiO_4)_2$	$[SiO_4]$, $[O_4H_4]$	[24]	[24]
Datolite	$CaBSiO_4(OH)$	$[SiO_4]$	[25]	[16]
Euclase	$BeAl(OH)SiO_4$	$[SiO_4]$	[26]	[14]
Chloritoide	$FeAl_2(OH)_2OSiO_4$	$[SiO_4]$	[27]	[28]
Staurolite	$Fe_4Al_{18}Si_8O_{46}(OH)_2$	$[SiO_4]$	[29]	[30]
Lawsonite	$CaAl_2Si_2O_7(OH)_2 \cdot H_2O$	$[Si_2O_7]$	[9]	[10]
Hennomartinite	$SrMn_2Si_2O_7(OH)_2 \cdot H_2O$	$[Si_2O_7]$	[31]	[10]
Hemimorphite	$Zn_4Si_2O_7(OH)_2 \cdot H_2O$	$[Si_2O_7]$	[12]	[32]
Dehyd. Hemim.	$Zn_4Si_2O_7(OH)_2$	$[Si_2O_7]$	[33]	[33]
Ilvaite	$CaFe_2Fe(OH)OSi_2O_7$	$[Si_2O_7]$	[34]	[35]
Axinite	$Ca_2FeAl_2BO(OH)(Si_2O_7)_2$	$[Si_2O_7]$	[36]	[37]
Bertrandite	$Be_4(OH)_2Si_2O_7$	$[Si_2O_7]$	[38]	[14]
Epidote	$Ca_2(Al,Fe)_3O(OH)Si_2O_7SiO_4$	$[Si_2O_7]$, $[SiO_4]$	[39]	[16]
Clinozoisite	$Ca_2Al_3O(OH)Si_2O_7SiO_4$	$[Si_2O_7]$, $[SiO_4]$	[40]	[16]
Zoisite	$Ca_2Al_3O(OH)Si_2O_7SiO_4$	$[Si_2O_7]$, $[SiO_4]$	[40]	[16]
Pectolite	$Ca_2NaSi_3O_8(OH)$	$[Si_3O_8(OH)]$ chain	[41]	[42]
Serandite	$Mn_2NaSi_3O_8(OH)$	$[Si_3O_8(OH)]$ chain	[43]	[42]
Bazzite	$Be_3Sc_2Si_6O_{18} \cdot xH_2O,Na$	$[Si_6O_{18}]$ ring	[44]	[44]
Beryl	$Be_3Al_2Si_6O_{18} \cdot xH_2O,Na$	$[Si_6O_{18}]$ ring	[45]	[46]
Cordierite	$Mg_2Al_4Si_5O_{18} \cdot xH_2O,Na$	$[Al_2Si_4O_{18}]$ ring	[47]	[48]
Armenite	$BaCa_2Al_6Si_9O_{30} \cdot 2H_2O$	$[Al_3Si_9O_{30}]$ ring	[49]	[49]
Milarite	$KCa_2AlBe_2Si_{12}O_{30} \cdot xH_2O$	$[Si_{12}O_{30}]$ ring	[49]	[49]
Tourmaline	$NaMg_3Al_6(BO_3)_3(OH)_4Si_6O_{18}$	$[Si_6O_{18}]$ ring	[50]	[51]
Tremolite	$Ca_2Mg_5(OH)_2Si_8O_{22}$	$[Si_8O_{22}]$ ribbon	[52]	[53]
Fe-Aktinolite	$Ca_2(Fe,Mg)_5(OH)_2Si_8O_{22}$	$[Si_8O_{22}]$ ribbon	[54]	[55]
Talc	$Mg_3(OH)_2Si_4O_{10}$	$[Si_4O_{10}]$ layer	[56]	[16]
Pyrophyllite	$Al_2(OH)_2Si_4O_{10}$	$[Si_4O_{10}]$ layer	[57]	[16]
Lizardite	$Mg_3(OH)_4Si_2O_5$	$[Si_2O_5]$ layer	[58]	[16]
Kaolinite	$Al_2(OH)_4Si_2O_5$	$[Si_2O_5]$ layer	[59]	[16]
Muscovite	$KAl_2(OH)_2Si_3AlO_{10}$	$[Si_3AlO_{10}]$ layer	[60]	[16]
Phlogopite	$KMg_3(OH)_2Si_3AlO_{10}$	$[Si_3AlO_{10}]$ layer	[61]	[16]
Annite	$KFe_3(OH)_2Si_3AlO_{10}$	$[Si_3AlO_{10}]$ layer	[62]	[63]
Analcime	$NaAlSi_2O_6 \cdot H_2O$	$[(Al,Si)_nO_{2n}]$	[64]	[65]
Natrolite	$Na_2Al_2Si_3O_{10} \cdot 2H_2O$	$[(Al,Si)_nO_{2n}]$	[66]	[16]
Phase A	$Mg_7Si_2O_8(OH)_6$	$[SiO_4]$	[67]	[67]
Phase B	$Mg_{12}Si_4O_{19}(OH)_2$	$[SiO_6]$, $[SiO_4]$	[67]	[67]
Goethite	$\alpha\text{-}FeOOH$	$[FeO_3(OH)_3]$	[68]	[69]
Lepidocrocite	$\gamma\text{-}FeOOH$	$[FeO_4(OH)_2]$	[14]	[14]
Diaspore	$\alpha\text{-}AlOOH$	$[AlO_3(OH)_3]$	[70]	[71]
Boehmite	$\gamma\text{-}AlOOH$	$[AlO_4(OH)_2]$	[72]	[14]
Groutite	$\alpha\text{-}MnOOH$	$[MnO_3(OH)_3]$	[71]	[71]
Manganite	$\gamma\text{-}MnOOH$	$[MnO_3(OH)_3]$	[71]	[71]
Brucite	$Mg(OH)_2$	$[Mg(OH)_6]$	[73]	[14]

Table 1. (*continued*)

Mineral	Formula	Structural elements	Structure	IR data
Gibbsite	$Al(OH)_3$	$[Al(OH)_6]$	[74]	[14]
Trona	$Na_3(CO_3HCO_3) \cdot 2H_2O$	$[CO_3]$, $[CO_2(OH)]$	[75]	[15]
Azurite	$Cu_3(OH)_2(CO_3)_2$	$[CO_3]$	[76]	[18]
Malachite	$Cu_2(OH)_2CO_3$	$[CO_3]$	[77]	[18]
Natrochalcite	$(Na,K)Cu_2(S,Se)O_4 \cdot H_3O_2$	$[(S,Se)O_4]$	[78]	[79]
Antlerite	$Cu_3(OH)_4SO_4$	$[SO_4]$	[80]	[13]
Brochantite	$Cu_4(OH)_6SO_4$	$[SO_4]$	[81]	[18]
Jarosite	$KFe_3(OH)_6(SO_4)_2$	$[SO_4]$	[82]	[13]
Alunite	$KAl_3(OH)_6(SO_4)_2$	$[SO_4]$	[82]	[16]
Natroalunite	$NaAl_3(OH)_6(SO_4)_2$	$[SO_4]$	[83]	[16]
Polyhalite	$K_2Ca_2Mg(SO_4)_4 \cdot 2H_2O$	$[SO_4]$	[84]	[13]
Gypsum	$CaSO_4 \cdot 2H_2O$	$[SO_4]$	[85]	[16]
OH-Apatite	$Ca_5(OH)(PO_4)_3$	$[PO_4]$	[86]	[87]
Lazulite	$(Mg,Fe)Al_2(OH)_2(PO_4)_2$	$[PO_4]$	[88]	[13]
Turquoise	$CuAl_6(OH)_8(PO_4)_4 \cdot 4H_2O$	$[PO_4]$	[89]	[13]
Pseudomalachit	$Cu_5(OH)_4(PO_4)_2$	$[PO_4]$	[90]	[13]
Variscite	$AlPO_4 \cdot 2H_2O$	$[PO_4]$	[91]	[92]
Scorodite	$FeAsO_4 \cdot 2H_2O$	$[AsO_4]$	[93]	[13]
Olivenite	$Cu_2(OH)AsO_4$	$[AsO_4]$	[94]	[13]

Table 2. Compilation of distance and frequency ranges for the three hydrous species H_2O, $H_3O_2^-$, and OH^-

Hydrous species	O···O Distances/Å	Stretching frequencies/cm^{-1}
H_2O	2.59–3.48	2838–3663
$H_3O_2^-$	2.44–2.48[a]	~1200[a]
OH^-	2.46–3.69	1000–3738

[a] Values for the internal pseudosymmetric H bond of the molecule

OH^-, H_2O, or even $H_3O_2^-$ groups. Table 2 compiles the distance and frequency ranges of these three groups. Thus, H_2O molecules show rather long H bonds and high frequencies, whereas $H_3O_2^-$ units display very low frequencies due to very short H bonds. In contrast, hydroxyl groups are distributed from very strong to very weak H bonds and from very low to very high stretching frequencies.

Results and Discussion

The correlation of $d(O···O)$ *vs.* frequency is plotted in Fig. 1. A data sheet with all relevant distance and wavenumber data can be obtained from the author upon request. The 125 data pairs cover a wavenumber range of 1000 to 3738 cm^{-1} and an O···O distance range of 2.44 to 3.5 Å. Thus, they cover H bonds from very short, almost symmetric ones to very weak H bonds or almost unbonded entities. The plot shows an empty data region between 1500 and 2650 cm^{-1}. This apparently wide gap, however, corresponds to only 2.50 to 2.58 Å in terms of

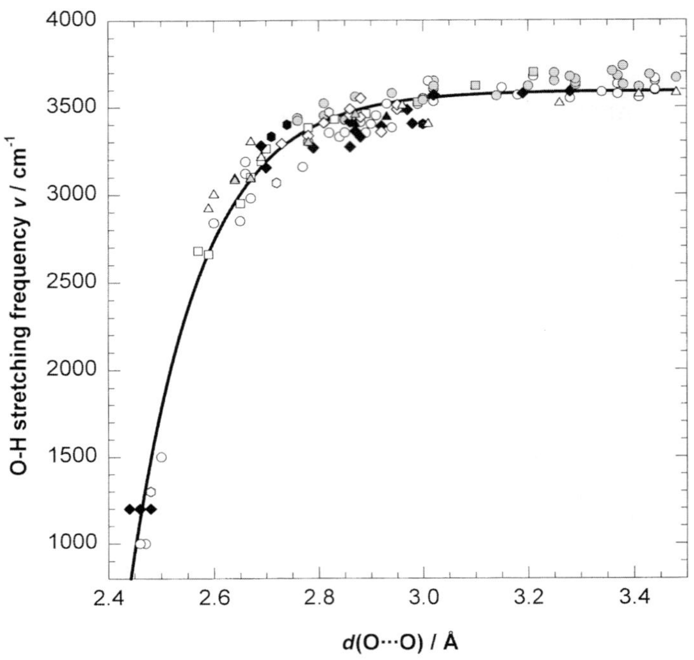

Fig. 1. Plot of the $d(\text{O}\cdots\text{O})$ – frequency correlation; open symbols represent straight H bonds, shaded symbols mark bent H bonds, and filled ones denote copper compounds; circles – silicates, squares - (oxy)hydroxides, hexagons – carbonates, diamonds – sulfates, triangles – phosphates and arsenates; the regression curve was calculated for distances < 3.5 Å ($n = 124$) in the form $v = 3592-$
$$304 \cdot 10^9 \cdot \ \exp(-d/0.1321), \ R^2 = 0.96$$

distance. A remote datapoint (a copper arsenate) at $d(\text{O}\cdots\text{O}) = 3.69$ Å (which is evidently far from H bonding and not visible in Fig. 1) was included because of its unusually low stretching frequency of only $3400\,\text{cm}^{-1}$. This feature, as well as the generally wide scatter of data points from copper compounds (solid symbols in Fig. 1) will be discussed below.

The correlation of $d(\text{H}\cdots\text{O})$ *vs.* stretching frequency is given in Fig. 2. Because of the above mentioned limitations and lack of sufficiently accurate data, only 47 data pairs are included in the correlation. The low frequency part, representing very strong H bonds, is rather underrepresented and shows extreme scatter. As a matter of fact, the empty wavenumber region is also observed in Fig. 2, and corresponds to a distance range of $\sim$1.45 to 1.60 Å (if the dashed extension line of the regression function is considered representative).

The data of both plots were fitted by a regression function in the form: $v = A - B \cdot \exp \cdot (-d/C)$. A similar function had been employed successfully in a previous correlation [7]. The units of v, A, and B are cm^{-1}, whereas d and C are expressed in Å. A list of resulting regression parameters for all data pairs and for various subgroups is presented in Table 3. For the $d(\text{O}\cdots\text{O})$-$v$ correlation the resulting regression coefficients R^2 are better than 0.96 except for the subgroup without the seven very strong H bond data (> 2.55 Å, $> 2000\,\text{cm}^{-1}$) where R^2 amounts to only 0.84. It is increased towards 0.98 if the data are restricted in terms of geometry (only straight bonds) or in terms of chemistry (only silicates; no copper

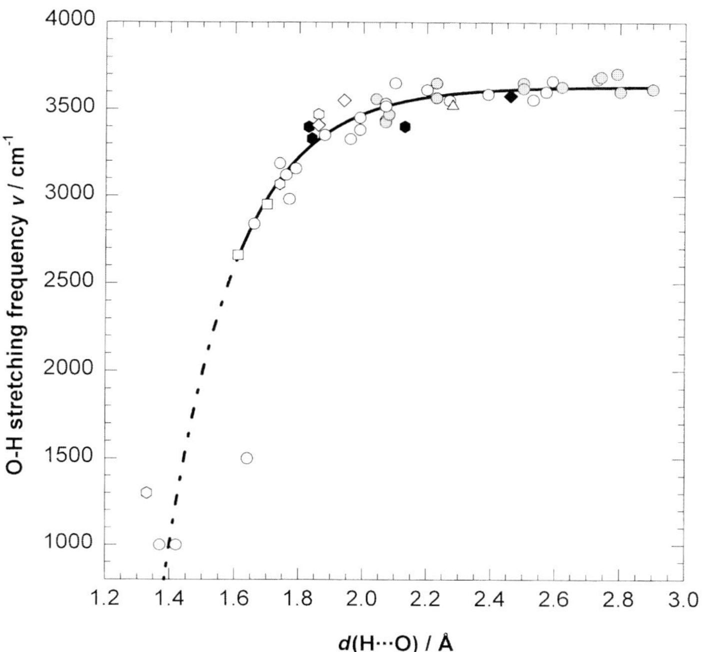

Fig. 2. Plot of the d(H$\cdots$O) – frequency correlation; symbol code as in Fig. 1; the regression curve was calculated for data above $2500\,\mathrm{cm}^{-1}$ ($n = 43$) in the form $v = 3632 - 1.79 \cdot 10^6 \cdot \exp(-d/0.2146)$, $R^2 = 0.91$, and was extrapolated towards lower frequencies (dashed curve)

Table 3. Parameters of the regression function $v = A - B \cdot \exp(-d/C)$ for all data and for different subgroups; a: d(O$\cdots$O)-v correlation; b: d(H$\cdots$O)-v correlation

a:

Dataset and constraints	n	A/cm^{-1}	B/cm^{-1}	$C/\mathrm{Å}$	R^2
All data	125	3589	$331 \cdot 10^9$	0.1315	0.96
$d < 3.5\,\mathrm{Å}^{\,a}$	124	3592	$304 \cdot 10^9$	0.1321	0.96
$d \leq 3.2\,\mathrm{Å}$	95	3545	$1.19 \cdot 10^{12}$	0.1230	0.96
$2.55\,\mathrm{Å} < d \leq 3.2\,\mathrm{Å}$	88	3589	$2.97 \cdot 10^9$	0.1706	0.84
Only straight H bonds	80	3554	$635 \cdot 10^9$	0.1270	0.97
Only silicates	64	3622	$238 \cdot 10^9$	0.1346	0.98
Without Cu; $d \leq 3.2\,\mathrm{Å}$	75	3530	$17.9 \cdot 10^{12}$	0.1088	0.97

b:

Dataset and constraints	n	A/cm^{-1}	B/cm^{-1}	$C/\mathrm{Å}$	R^2
All data	47	3720	$257 \cdot 10^3$	0.2978	0.88
Without $v = 1500\,\mathrm{cm}^{-1}$	46	3680	$411 \cdot 10^3$	0.2683	0.95
$v > 2500\,\mathrm{cm}^{-1\,a}$	43	3632	$1.79 \cdot 10^6$	0.2146	0.91
$v > 2500\,\mathrm{cm}^{-1}$; $d < 2.3\,\mathrm{Å}$	30	3599	$4.00 \cdot 10^6$	0.1935	0.89
$v > 2500\,\mathrm{cm}^{-1}$; without Cu	39	3642	$1.38 \cdot 10^6$	0.2227	0.93

[a] Regression curve plotted in Figs. 1 and 2

compounds). For the $d(\mathrm{H}\cdots\mathrm{O})$-$v$ correlation R^2 approaches 0.95 if the data point at $v = 1500\,\mathrm{cm}^{-1}$ (probably an outlier) is omitted, whereas it is only 0.88 for all 47 data pairs included.

Reasons for scatter of data

A short inspection of Figs. 1 and 2 reveals that deviations from the regression lines may amount to $150\,\mathrm{cm}^{-1}$ or even more in some cases. Extreme outliers (*e.g.* among the low-energy data of very strong H bonds in Fig. 2) are distinguished from systematic deviations (*i.e.* data of bent H bonds in Fig. 1), and random scattering that may be strong in certain compounds (*e.g.* Cu minerals) and which is attributed to effects different from H bonding (*e.g.* cationic effects, factor group splitting, dynamic proton behavior).

The most evident outliers occur in Fig. 2 where four data points between 1000 and $1500\,\mathrm{cm}^{-1}$ correlate to varying $\mathrm{H}\cdots\mathrm{O}$ distances between 1.33 and 1.64 Å, even if the expected distance interval (according to the steep slope of the regression line) should be rather narrow. Figure 1 confirms the low wavenumbers, hence the proton-acceptor bond lengths in Fig. 2 must be rather inaccurate. In spite of the quality criteria used (*i.e.* only O–H > 0.90 Å), a closer inspection of data reveals that the O–H values from the structure refinements are distorted in some cases. For example, the most severe outlier at $1500\,\mathrm{cm}^{-1}$, mozartite [19], passed the quality check with an O–H distance of 0.91 Å (from X-ray data), and thus yielded $d(\mathrm{H}\cdots\mathrm{O}) = 1.61$ Å. According to a previous $d(\mathrm{O-H})$-v diagram [6], at a wavenumber of $1500\,\mathrm{cm}^{-1}$ an O–H distance of $\sim$1.15 Å is expected. Thus, it is likely that the true data point should be shifted by approximately 0.2 Å to the left. In general, data on very strong H bonds may commonly contain deviations, either in terms of frequency (because of the extremely broad stretching bands) or in terms of proton positions (because of difficulties to refine pseudosymmetric proton sites even with neutrons) [79].

Bent H bonds which yield constantly too high wavenumbers in the $d(\mathrm{O}\cdots\mathrm{O})$-$v$ plot (shaded symbols in Fig. 1) are the most important source of systematic deviations. However, this observation is quite reasonable and in good agreement with considerations on bond geometry. The attractive influence of the H bond acceptor is attenuated if the $\mathrm{H}\cdots\mathrm{O}$ distance in a bent H bond is longer (as a consequence of the bent geometry) than in a straight H bond. As a logical consequence, the $d(\mathrm{H}\cdots\mathrm{O})$-$v$ plot in Fig. 2 does not show this systematic scatter towards higher wavenumbers. A sample case is dehydrated hemimorphite [33], where the stretching frequencies of the straight ($3531\,\mathrm{cm}^{-1}$) and bent H bonds ($3603\,\mathrm{cm}^{-1}$) belong to the same $\mathrm{O}\cdots\mathrm{O}$ distance of 3.02 Å. The data point for the former is almost exactly on the regression line, whereas the latter is found above the curve in Fig. 1. In contrast, the respective $\mathrm{H}\cdots\mathrm{O}$ distances, *i.e.* 2.07 Å for the former and 2.80 Å for the latter, are strongly different and thus plot correctly on the regression line in Fig. 2. A different situation is observed in case of bifurcated bonds, where in spite of bent bonds the data point is found at or below the regression line due to the doubled attractive force of the two proton acceptors, *e.g.* datolite, represented by the almost hidden datapoint at 2.96 Å /$3493\,\mathrm{cm}^{-1}$ in Fig. 1.

Even though H bonding is extremely weak or absent beyond 3.2 Å (Fig. 1) or 2.3 Å (Fig. 2), data show still appreciable scatter. Hence, these deviations must be caused by effects which are not related to H bonding. Different cationic influences are among the most prominent ones. Starting from the free OH^- ion, stretching frequencies increase by more than $100\,cm^{-1}$ with decreasing metal-oxygen distances in the coordination sphere of the ion [95]. In addition, the hydrogen bond donor strength increases. This synergetic effect is especially strong for copper ions in the coordination sphere of the hydroxyl group [5]. The effect is reversed, if the H bond acceptor is closely coordinated by these metal ions, *e.g.* in malachite, azurite, and brochantite [18]. In general, these effects seem to be the reason for the strong scatter among the copper compounds (solid symbols in Figs. 1 and 2) and even among data of a single mineral (brochantite). The unusually low frequency of $3440\,cm^{-1}$ (without any H bonding) in olivenite appears to be caused also by the peculiar role of copper in the close neighbourhood of the hydroxyl group. Metal ions with different electronegativities in the coordination sphere of the OH^- ion also influence the band positions. This was demonstrated in talc samples ($v = 3678\,cm^{-1}$) where Mg was replaced successively by transition metal ions which resulted in a maximum frequency decrease by $50\,cm^{-1}$ for the fully exchanged endmembers [95]. Similar shifts were observed in the amphibole group changing from pure tremolite (Mg) to richterites (K, Fe) [53] and actinolites (Fe) [55].

Factor group splitting (*i.e.* coupling of vibrations in the primitive unit cell) of bands may be another reason for scattered data points. For example, boehmite shows two bands at 3087 and $3283\,cm^{-1}$, whereas the isotopically diluted sample shows only a single band at $3180\,cm^{-1}$ [14]. However, as this rather extreme case demonstrates, vibrational coupling leads to data scattering but not to systematic shifts, as the mean of both coupled modes is quite similar to the uncoupled vibration.

Finally, scatter observed in Fig. 2 may be widely caused by proton dynamics. If O–H vectors show librational or hopping motions, the respective O–H bonds may appear considerably shortened by diffraction methods [22]. For the same reason a d(O–H)-v correlation was not presented in this paper.

Differences in regression functions

The parameter A of the regression function represents the upper limiting O–H stretching frequency if the H bond acceptor is missing or extremely remote from the proton. Inspection of Table 3a yields A values from 3530 to $3622\,cm^{-1}$ depending upon the dataset constraints. It must be emphasized that predominantly datapoints from bent H bonds which are (almost) not engaged in H bonds (> 3.2 Å), *e.g.* layer and ribbon silicates, display rather high wavenumbers in the upper right corner of Fig. 1. If these distorted values are removed (constraints: "$d \leq 3.2$ Å"; "only straight H bonds"; "without Cu and $d \leq 3.2$ Å"), a limiting wavenumber A between 3530 and $3554\,cm^{-1}$ is approached, which is in excellent agreement with theoretical and experimental values for free OH^- ions, *i.e.* 3554–$3562\,cm^{-1}$ [5]. The highest A value is shown by the "only silicates" subgroup, which is easily explained by the impact of the above mentioned layer and ribbon silicates. The

Table 4. Comparison of calculated data points from regression lines of the present and previous correlation diagrams; distances in Å, wavenumbers in cm^{-1}

d(O$\cdots$O)	Nakamoto [4]	Novak [6]	Mikenda [7]	This study[a]	This study[b]
2.5	2070	1500	–	1758	–
2.6	2500	2680	2880[c]	2732	2870
2.7	2930	3200	3180	3188	3181
2.8	3320	3400	3345	3402	3352
2.9	3490	3500	3450	3503	3447
3.0	3600	–	3510	3550	3499
3.1	3630	–	3550[c]	3572	3528
3.2	3660	–	–	3582	–

[a] All data <3.5 Å ($n=124$); [b] only straight H bonds, 2.55 Å $<d\leq3.2$ Å ($n=59$); [c] values slightly outside the data range

curvature of the regression function is expressed by parameter B in the way that high values indicate strong curvature (*i.e.* the subgroup "without Cu") and low values indicate only slight bending (*i.e.* the subgroup "2.55 Å $<d\leq3.2$ Å which resembles that of *Mikenda*'s plot [7]).

Figure 2 also contains the high frequency data points in the upper right corner and thus leads to rather elevated A parameters in Table 3b. In addition, due to the strong scatter of the four low frequency data points, regression functions are least-squares fitted only for the part >2500 cm^{-1} in three cases. Hence, even if H bonds are limited to distances <2.3 Å, a rather high A value of 3599 cm^{-1} is observed, which resembles that of Table 3a for the region "2.55 Å $<d\leq3.2$ Å. The B values, which are lower than in Table 3a, indicate reduced curvature.

Comparison to previous correlation diagrams

To evaluate the present distance-frequency correlation in minerals, the resulting regression lines are compared to published correlation trends. Because of a missing regression function in previous correlations [4, 6] or incompatibility (O–D vibrations in [7]), calculated data pairs across the whole range of H bonds are compiled in Table 4.

Compared to the diagram of *Nakamoto* [4], the present data exceed the old diagram in the low frequency region even if the d(O$\cdots$O) range is equivalent. Hence, the frequency at 2.5 Å is too high by ~300 cm^{-1}, whereas even the value of 1758 cm^{-1} of the present study appears high compared to the data points used in this region. The high frequency end of *Nakamoto*'s correlation line is also too high by ~100 cm^{-1}, because he assumed a value close to 3700 cm^{-1} for a free OH$^-$ ion. However, his curve is less bent than the present one and so it crosscuts the present line around 2300 and 3500 cm^{-1}. Between these intersections, *Nakamoto*'s data appear too low by more than 200 cm^{-1} (at 2.6 Å). In general, it is a handicap of this previous correlation that it was established from only 26 data points.

Comparison of the present data to the more recent diagram of *Novak* [6] gives excellent agreement between the regression curves. Only the low frequent end of the curve appears too high in the present study (as mentioned above) and might be better represented by *Novak*'s $1500\,\mathrm{cm}^{-1}$ at $2.5\,\mathring{\mathrm{A}}$. However, his diagram suffers from data restriction to distances below $2.9\,\mathring{\mathrm{A}}$. The scatter of data appears similar in both correlations, even if *Novak* used only half the number of data of the present correlation curve.

Comparison to the most recent distance-frequency correlation of *Mikenda* [7], who used only high quality data of isotopically diluted solid hydrates, also yields excellent agreement (deviation $\leq 50\,\mathrm{cm}^{-1}$). Only *Mikenda*'s data point at $2.6\,\mathring{\mathrm{A}}$ is too high by $\sim 150\,\mathrm{cm}^{-1}$. This is a consequence of the curvature of his trend which is rather flat compared to the present regression line. However, the O$\cdots$O distance of $2.6\,\mathring{\mathrm{A}}$ actually exceeds the data used by this author. To compare further, it appeared challenging to select a high quality data set from the present mineral data that resembles *Mikenda*'s data quite closely. Hence, the distance range was restricted to values between 2.55 and $3.2\,\mathring{\mathrm{A}}$, and only straight H bonds were used. Table 4 shows that the agreement between the regression lines of these selected data sets is perfect. Deviations amount to only $\sim 10\,\mathrm{cm}^{-1}$ except for the point at $3.1\,\mathring{\mathrm{A}}$ (actually outside *Mikenda*'s data [7]), where *Mikenda*'s curve is higher by $\sim 20\,\mathrm{cm}^{-1}$. This may be caused by the considerably higher A value ($\sim 3650\,\mathrm{cm}^{-1}$, converted from O–D to O–H wavenumbers) that was constrained to the frequency of a free HOD molecule in Ref [7].

In general, the more recent distance-frequency correlations [6, 7] are confirmed, and also the trend of a rather old study [4] is met. However, even if full agreement between a restricted dataset of the present study and the high quality data of Mikenda [7] may be obtained, the author is convinced that the complete range of data covering all H bond lengths and frequencies should be used for a true correlation. In contrast to *Mikenda*, the present regression lines as well as that of *Novak* [6] show a stronger curvature to meet the data of very strong H bonds. Moreover, according to visual inspection, a regression line with stronger curvature were easily fitted through *Mikenda*'s data (after release of the constrained A parameter) and thus confirms the present distance-frequency correlation in minerals.

Acknowledgements

The critical comments of *A. Beran* (Vienna), *H. Falk* (Linz), and *W. Mikenda* (Vienna) helped to improve the quality of the present paper.

References

[1] Emsley J (1980) Chem Soc Rev **9**: 91
[2] Rundle RE, Parasol M (1952) J Chem Phys **20**: 1487
[3] Bellamy LJ, Owen AJ (1969) Spectrochim Acta **A25**: 329
[4] Nakamoto K, Margoshes M, Rundle RE (1955) J Am Chem Soc **77**: 6480
[5] Lutz HD (1995) Struct Bond **82**: 85
[6] Novak A (1974) Struct Bond **18**: 177
[7] Mikenda W (1986) J Mol Struct **147**: 1
[8] Bell DR, Rossman GR (1992) Contrib Mineral Petrol **111**: 161
[9] Libowitzky E, Armbruster T (1995) Am Mineral **80**: 1277
[10] Libowitzky E, Rossman GR (1996) Am Mineral **81**: 1080

[11] ICSD 98/1 Inorganic Crystal Structure Database, FIZ Karlsruhe – Gmelin Institute
[12] Libowitzky E, Schultz AJ, Young DM (1998) Z Kristallogr **213**: 659
[13] Moenke H (1966) Mineralspektren II. Akademie-Verlag, Berlin
[14] Farmer VC (ed) (1974) The Infrared Spectra of Minerals. Miner Soc, London
[15] Jones GC, Jackson B (1993) Infrared Transmission Spectra of Carbonate Minerals. Chapman & Hall, London
[16] Salisbury JW, Walter LS, Vergo N, D'Aria DM (1992) Infrared (2.1–25 μm) Spectra of Minerals. Johns Hopkins, London
[17] Libowitzky E, Rossman GR (1996) Phys Chem Minerals **23**: 319
[18] Schmidt M, Lutz HD (1993) Phys Chem Minerals **20**: 27
[19] Nyfeler D, Hoffmann C, Armbruster T, Kunz M, Libowitzky E (1997) Am Mineral **82**: 841
[20] Zobetz E, Zemann J, Heger G, Voellenkle H (1979) Anz Österr Akad Wiss, Math-Naturwiss Kl **116**: 145
[21] Shinoda K, Aikawa N (1994) Phys Chem Minerals **21**: 24
[22] Lager GA, Armbruster T, Faber J (1987) Am Mineral **72:** 756
[23] Rossman GR, Aines RD (1991) Am Mineral **76**: 1153
[24] Armbruster T, Libowitzky E, Kunz M, Miletich R, Gutzmer J (1999) Am Mineral **84** (in preparation)
[25] Foit FF Jr, Phillips MW, Gibbs GV (1973) Am Mineral **58**: 909
[26] Hazen RM, Au AY, Finger LW (1986) Am Mineral **71**: 977
[27] Hanscom RH (1975) Acta Cryst **B31**: 780
[28] Fransolet A-M (1978) Bull Minéral **101**: 548
[29] Ståhl K, Kvick Å, Smith JV (1988) J Sol State Chem **73**: 362
[30] Koch-Müller M, Langer K, Beran A (1995) Phys Chem Minerals **22**: 108
[31] Libowitzky E, Armbruster T (1996) Am Mineral **81**: 9
[32] Libowitzky E, Rossman GR (1997) Eur J Mineral **9**: 793
[33] Libowitzky E, Kohler T, Armbruster T, Rossman GR (1997) Eur J Mineral **9**: 803
[34] Ghose S, Hewat AW, Marezio M (1984) Phys Chem Minerals **11**: 67
[35] Beran A, Bittner H (1974) Tscherm Min Petr Mitt **21**: 11
[36] Swinnea JS, Steinfink H, Rendon Diaz Miron LE, de la Vega SE (1981) Am Mineral **66**: 428
[37] Beran A (1971) Tscherm Min Petr Mitt **16**: 281
[38] Downs JW, Ross FK (1987) Am Mineral **72**: 979
[39] Nozik YZ, Kanepit VN, Fykin LY, Makarov YS (1978) Geochem Internat **15**: 66
[40] Comodi P, Zanazzi PF (1997) Am Mineral **82**: 61
[41] Takéuchi Y, Kudoh Y (1977) Z Kristallogr **146**: 281
[42] Hammer VMF, Libowitzky E, Rossman GR (1998) Am Mineral **83**: 569
[43] Jacobsen SD (1998) MS Thesis, University of Colorado, Boulder
[44] Armbruster T, Libowitzky E, Diamond L, Auernhammer M, Bauerhansl P, Hoffmann C, Irran E, Kurka A, Rosenstingl H (1995) Miner Petrol **52**: 113
[45] Artioli G, Rinaldi R, Ståhl K, Zanazzi PF (1993) Am Mineral **78**: 762
[46] Aurisicchio C, Grubessi O, Zecchini P (1994) Can Mineral **32**: 55
[47] Wallace JH, Wenk HR (1980) Am Mineral **65**: 96
[48] Aines RD, Rossman GR (1984) Am Mineral **69**: 319
[49] Armbruster T (1999) Am Mineral **84**: 92
[50] Hawthorne FC, MacDonald DJ, Burns PC (1993) Am Mineral **78**: 265
[51] Gonzalez-Carreño T, Fernández M, Sanz J (1988) Phys Chem Minerals **15**: 452
[52] Yang H-X, Evans BW (1996) Am Mineral **81**: 1117
[53] Gottschalk M, Andrut M (1998) Phys Chem Minerals **25**: 101
[54] Urusov VS, Zver'kova ON, Yamnova NA, Polosin AV (1987) Vestnik Moskovskogo Universiteta, Geologiya **1987**: 43
[55] Skogby H, Rossman G (1991) Phys Chem Minerals **18**: 64

[56] Perdikatsis B, Burzlaff H (1981) Z Kristallogr **156**: 177

[57] Lee JH, Guggenheim S (1981) Am Mineral **66**: 350

[58] Gregorkiewitz M, Lebech B, Mellini M, Viti C (1996) Am Mineral **81**: 1111

[59] El Sayed K, Heiba ZK, Abdel Rahman AM (1990) Cryst Res Tech **25**: 305

[60] Catti M, Ferraris G, Hull S, Pavese A (1994) Eur J Mineral **6**: 171

[61] Rayner JH (1974) Min Mag **39**: 850

[62] Hazen RM, Burnham CW (1973) Am Mineral **58**: 889

[63] Redhammer GJ, Beran A, Dachs E, Amthauer G (1993) Phys Chem Minerals **20**: 382

[64] Ferraris G, Jones DW, Yerkess J (1972) Z Kristallogr **135**: 240

[65] Libowitzky E, Rossman GR (1997) Am Mineral **82**: 1111

[66] Stuckenschmidt E, Joswig W, Baur WH (1993) Phys Chem Minerals **19**: 562

[67] Hofmeister AM, Cynn H, Burnley PC, Meade C (1999) Am Mineral **84**: 454

[68] Szytula A, Burewicz A, Dimitrijewic Z, Krasnicki S, Rzany H, Todorovic J, Wanic A, Wolski W (1968) Phys Stat Sol **26**: 429

[69] Libowitzky E (1996) Mitt Österr Miner Ges **141**: 134

[70] Hill RJ (1979) Phys Chem Minerals **5**: 179

[71] Kohler T, Armbruster T, Libowitzky E (1997) J Solid State Chem **133**: 486

[72] Christoph GG, Corbato CE, Hofmann A, Tettenhorst RT (1979) Clays Clay Miner **27**: 81

[73] Zigan F, Rothbauer R (1967) N Jb Miner Mh **1967**: 137

[74] Saalfeld H, Wedde M (1974) Z Kristallogr **139**: 129

[75] Pertlik F (1986) Mitt Österr Miner Ges **131**: 7

[76] Zigan F, Schuster HD (1972) Z Kristallogr **135**: 416

[77] Zigan F, Joswig W, Schuster HD (1977) Z Kristallogr **145**: 412

[78] Giester G (1989) Z Kristallogr **187**: 239

[79] Beran A, Giester G, Libowitzky E (1997) Mineral Petrol **61**: 223

[80] Hawthorne FC, Groat LE, Eby RK (1989) Can Mineral **27**: 205

[81] Helliwell M, Smith JV (1997) Acta Cryst **C53**: 1369

[82] Menchetti S, Sabelli C (1976) N Jb Miner Mh **1976**: 406

[83] Okada K, Hirabayashi J, Ossaka J (1982) N Jb Miner Mh **1982**: 534

[84] Schlatti M, Sahl K, Zemann A, Zemann J (1970) Tscherm Min Petr Mitt **14**: 75

[85] Pedersen BF, Semmingsen D (1982) Acta Cryst **B38**: 1074

[86] Hughes JM, Cameron M, Crowley KD (1989) Am Mineral **74**: 870

[87] Engel G, Klee WE (1972) J Solid State Chem **5**: 28

[88] Giuseppetti G, Tadini C (1983) N Jb Miner Mh **1983**: 410

[89] Cid-Dresdner H (1965) Z Kristallogr **121**: 87

[90] Shoemaker GL, Anderson JB, Kostiner E (1981) Am Mineral **66**: 169

[91] Kniep R, Mootz D, Vegas A (1977) Acta Cryst **B33**: 263

[92] Salvador Salvador P, Fayos J (1972) Am Mineral **57**: 36

[93] Hawthorne FC (1976) Acta Cryst **B32**: 2891

[94] Toman K (1977) Acta Cryst **B33**: 2628

[95] Beckenkamp K, Lutz HD (1992) J Mol Struct **270**: 393

Received November 9, 1998. Accepted (revised) December 8, 1998

SpringerChemistry

Fortschritte der Chemie organischer Naturstoffe / Progress in the Chemistry of Organic Natural Products

Founded by L. Zechmeister
Editors: W. Herz, H. Falk, G. W. Kirby, R. E. Moore, C. Tamm (eds.)

Volume 78

1999. VIII, 168 pages. 2 figures.
Hardcover DM 250,–, öS 1750,–. Special price
for subscribers to the series: DM 225,–, öS 1575,–
ISBN 3-211-83311-0

Contents
Brassinosteroids (G. Adam, J. Schmidt, B. Schneider):
Introduction • Natural Occurrence and Distribution • Structures • Isolation
and Purification • Analysis of Brassinosteroids • Synthesis; Biosynthesis; Metabolism of Brassinosteroids • Physiological Action • Molecular Mode of Action
• Conclusions • References

Chemistry of the Neem Tree (Azadirachta indica A. Juss.) (A. Akhila, K. Rani):
Introduction • Chemistry of Limonoids • Other Compounds • References

Volume 77

1999. VII, 187 pages. 3 partly coloured figures.
Hardcover DM 250,–, öS 1750,–. Special price
for subscribers to the series: DM 225,–, öS 1575,–
ISBN 3-211-83264-5

Contents
Secondary Metabolites and the Control of Some Blue Stain and Decay Fungi
(W. A. Ayer, L. S. Trifonov)
Condensed Tannins (D. Ferreira, E. V. Brandt, J. Coetzee, E. Malan)
Constituents of *Lactarius* (Mushrooms) (W. M. Daniewski, G. Vidari)

All prices are recommended retail prices

SpringerWienNewYork

Sachsenplatz 4–6, P.O.Box 89, A-1201 Wien, Fax +43-1-330 24 26, e-mail: books@springer.at, Internet: http://www.springer.at
New York, NY 10010, 175 Fifth Avenue • D-14197 Berlin, Heidelberger Platz 3 • Tokyo 113, 3–13, Hongo 3-chome, Bunkyo-ku

Springer-Verlag
and the Environment

WE AT SPRINGER-VERLAG FIRMLY BELIEVE THAT AN
international science publisher has a special obliga-
tion to the environment, and our corporate policies
consistently reflect this conviction.

WE ALSO EXPECT OUR BUSINESS PARTNERS – PRINTERS,
paper mills, packaging manufacturers, etc. – to commit
themselves to using environmentally friendly mate-
rials and production processes.

THE PAPER IN THIS BOOK IS MADE FROM NO-CHLORINE
pulp and is acid free, in conformance with international standards for paper permanency.